THE POEM ELECTRIC

Technology and the American Lyric

S ETH P ERLOW

U NIVERSITY OF M INNESOTA P RESS
M INNEAPOLIS • L ONDON

Published by the University of Minnesota Press
111 Third Avenue South, Suite 290
Minneapolis, MN 55401-2520
http://www.upress.umn.edu

Printed in the United States of America on acid-free paper

The University of Minnesota is an equal-opportunity educator and employer.

Library of Congress Cataloging-in-Publication Data
Names: Perlow, Seth, author.
Title: The poem electric : technology and the American lyric / Seth Perlow.
Description: Minneapolis : University of Minnesota Press, 2018. |
    Includes bibliographical references and index.
Identifiers: LCCN 2018016693 (print) | ISBN 978-1-5179-0365-7 (hc) |
    ISBN 978-1-5179-0366-4 (pb)
Subjects: LCSH: American poetry—20th century—History and criticism. | Experimental poetry, American—
    History and criticism. | Literature and technology. | Rationalism. | BISAC: LITERARY CRITICISM / Poetry. |
    SOCIAL SCIENCE / Poverty & Homelessness. | TECHNOLOGY & ENGINEERING / Social Aspects.
Classification: LCC PS309.E96 P47 2018 (print) | DDC 811/.009356—dc23
LC record available at https://lccn.loc.gov/2018016693

UMP BmB 2018

For my parents,
Joan and David Perlow

Contents

Introduction

Technologies of Lyric Exemption

A poem is a small (or large) machine made of words.

—William Carlos Williams, *The Wedge*

It is difficult
to get the news from poems
yet men die miserably every day
for lack
of what is found there.

—William Carlos Williams, "Asphodel, that Greeny Flower"

Together, these maxims William Carlos Williams wrote around the middle of the twentieth century encapsulate the guiding concern of this study. On one hand, his claim that poems do not deliver "the news" reflects a common belief that poetry does something other than give practical information—a belief as old as Western poetics. Few people agree about what makes poetry valuable, but in defining its value, surprisingly many draw this distinction between factual knowledge and "what is found" in poems. These lines from Williams and many like them produce a theory of poetic value by rhetorically differentiating poetry from information: whatever proves special about poetry has something to do with its difference from "the news." On the other hand, by equating a poem with a machine, Williams evokes a comparably long history of technological figures for poetry. Poetry can seem technological in various senses. In *The Wedge*, Williams imagines a poetic efficiency in which "there can be no part, as in any other machine, that is redundant."[1] Elsewhere, he uses electrical language for poetry's technological dimension, describing poetic "imagination as . . . an electricity."[2] The electrical analogy for poetry has ancient roots, but it gained currency as modern technoscience rendered electricity more familiar and useful.[3] In 1867, when Walt Whitman professed to

1

"Sing the Body Electric," he also observed "the electric telegraph, stretching across the Continent."[4] By 1944, when Williams equated poem and machine, the proliferation of consumer electronics had accelerated dramatically and the research disciplines now known as computer science had gained significant traction. Since then, electricity has steadily become a less esoteric, more concrete and familiar figure for poetic technology. The family of devices called information technologies has suffused the mediascape through which poetry circulates. To call a poem "electric" today invokes not a mysterious natural force but a series of quotidian, instrumental functions.

This study therefore investigates what happens when a literary genre often valued for its difference from information and knowledge begins to circulate through technologies predominantly viewed as informational and logical. Far from rationalizing poetry, electronics intensify the rhetorical exemptions that construct poetry's value in terms of its difference from knowledge. Those who address poetic technologies within the logical, informational frameworks of contemporary technoscience overlook the common belief that poetry opposes and evades precisely such knowledge-oriented frameworks.

In *The Poem Electric,* I identify a lineage of writers who use electronics to distinguish poetic thought from rationalism. These writers view electronics as capable of sustaining alternatives to the rational processes that such devices normally seem to support. This account differs from the common view of electronics as knowledge engines, "information technologies" that aid knowledge production.[5] Certainly some recent poetry emphasizes computers' informational functions. When scholars celebrate such work as fulfilling poetry's new technological destiny, however, they elide the persistent lyric norm of valorizing poetry's exemption from rational, informational purposes. They also minimize an important dimension of daily life with electronics: although logic strictly governs a computer's physical function, we often interact with such devices in less orderly, rational ways. When I procrastinate online, I am not looking for information, merely distraction and amusement; and when I have technical problems with a computer, I often endure confusion, disorientation, and frustration before finding a solution. Media theorists and technologists have long addressed electronics outside the informational, rationalistic frameworks that dominate today's discussions of literary technologies. Computer scientists working in wartime cybernetics labs already saw that the challenge of robust artificial intelligence lies not merely in mechanizing rationalism but in making a device capable of something like intuition, ambivalence, or even emotion. The writers discussed in this book use a variety of electronics, familiar and esoteric, to celebrate lyric poetry's exemption from rational thought, its power to disrupt the orderly protocols of critique. Neither valorizing nor disparaging such claims of lyric exemption, *The Poem Electric* argues that the electrification of American verse cultures has energized the rhetoric that distinguishes poetic thought from rational knowledge.

The rhetoric of lyric exemption constructs a poem's value by distinguishing it from rational pursuits of knowledge, however construed. The specific predicates of exemption, against which poetic value gets positioned, vary within a more or less coherent domain of synonyms for knowledge in technoscientific discourse: information, data, news, certainty, logic, rationality, and so on. The operative theory of knowledge and its attendant vocabulary vary from case to case, but the rhetorical gesture itself, which celebrates poetry by exempting it from one kind of knowledge or another, pervades discourses of poetic value. This gesture involves lyric poems in circuits of exchange with the modes of knowledge they oppose, enabling poets and their readers to reconsider what it can mean to know. As a study of this rhetorical move and its enabling technologies, *The Poem Electric* hopes to consolidate other recent scholarship that develops a searchingly diverse vocabulary for poetry's difference from knowledge. Books by Daniel Tiffany, Steve McCaffery, Virginia Jackson, and Craig Dworkin propose terms such as "infidelity," "darkness," "pathos," and "illegibility" to distinguish poetic thought from rationalism.[6] This study responds to such accounts by making explicit their shared rhetorical habit of exempting poetry from rational knowledge. I trace this rhetoric through a specific lineage of experimental writers, but their claims of exemption signal their continuity with a much longer, more diverse lyric tradition. This rhetoric of exemption is specifically lyrical, rather than broadly poetic, because the characteristics that differentiate poetry from rationalism are those most closely linked with the lyric in particular, such as its expressiveness, affective intensity, and ambiguity. Below I will linger with these terms for defining the lyric in order to address the lively debates they have recently occasioned. Not all poetry distances itself from information and knowledge. Some great poems in other genres, such as the epic or georgic, aim forthrightly to instruct and inform. But, within the Western lyric tradition whose afterlives this book explores, a striking variety of writers deploy a rhetoric of poetic value that exempts poetry from the strictures of rationalism.

Rhetorical claims against knowledge often collaborate with the modes of knowledge they disavow. By repudiating one kind of knowledge, we open the way for new claims of insight and certainty. In a heated argument, for instance, I might admit I cannot prove my point but still avow certainty because "I can just *feel* it." Such appeals are compelling because of emotion's difference from rationality, its immunity to skepticism. My feelings, qua feelings, enjoy an authority independent of logic or evidence. This rhetorical move separates emotion from rational knowing but also authorizes emotion as a basis for claims of certainty, putting knowledge and emotion into a system of exchange. As I argue below, a structurally equivalent rhetoric grounds the production of knowledge in the Cartesian philosophical tradition, where claims *not* to know enable new knowledge. By the same token, poets and their readers alternately celebrate poetry's exemption from rational knowledge and claim that it enables poems to sustain alternative ways of knowing. These gestures of disavowal are rhetorical, rather than nominalistic or metaphysical, first, because they exchange flexibly with the knowledge they disavow and, second, because their accordance with some preexisting epistemology matters less than their illocutionary force. Whether or not a poem really offers an alternative to rational, informational thinking, the technological supports and rhetorical effects of such claims shape how we write and read poetry.

Several key texts in this study move across the increasingly porous border between literary and scholarly writing, between poetry and criticism. Electronics enable us to read and write with the same equipment, technologically literalizing the indecision between creative and interpretive activities. The blurring of such generic and disciplinary distinctions leads experimentalists to write poetry that does the work of literary interpretation, criticism that is itself lyrical. These texts destabilize literary scholarship's broader relation to knowledge and the limits of its dedication to rationalism. If poems do not confine themselves to rational thought, then perhaps interpretations of poems should not either. Must humanities scholarship aim primarily to produce knowledge even if its objects of study resist such aims? An interpretive practice more faithful to poetry's exemption from rationalism might produce better readings. The chapters below discuss works of literary interpretation and poetics that complicate the assumption that literary criticism should yield rational knowledge. By opening such possibilities, the writers in this study bring American poetry into contact with some of the liveliest recent debates about the value of humanities scholarship. Although I do not practice such pararational modes of criticism in this book, I take them as a generative site of scholarly interest and an important component of the lyric's exemption from rationalism.

AFTERLIVES OF THE LYRIC

Distinctions between poetry and rationalism go back to antiquity. Plato saw writing as a technology that might impair our memory, and he assigned a dangerous ignorance to the poet, whose imitative art could mislead us.[7] Since the ascendancy of the modern lyric, however, the meaning of poetry's difference from rationalism has often changed. By the eighteenth century, classically inclined poets such as Thomas Gray had begun to transvalue poetry's exemption from knowledge-oriented discourse: "where ignorance is bliss / 'Tis folly to be wise."[8] The romantic poets may have shared little with Gray, but as they set the coordinates of the lyric, they made an axiom of Gray's poetic preference for ignorance. In "Intimations of Immortality" and elsewhere, Wordsworth extends Gray's affiliation of a nostalgically valued ignorance with a return to childhood: "Our birth is but a sleep and a forgetting . . . / Shades of the prison-house begin to close / Upon the growing Boy, / But He beholds the light, and whence it flows . . . / At length the Man perceives it die away, / And fade into the light of common day."[9] The equation of lyric inspiration with an impossible return to childlike naiveté reached American shores largely intact, as in Whitman: "A child said, *What is the grass?* . . . / How could I answer the child? I do not know what it is, any more than he."[10] A very long story could be told about the discourses of knowledge's others, poetic and philosophical, from antiquity to the present, a story that might include the negative theologies of the Middle Ages, for instance, and the speculative philosophies of today. As far as contemporary poetics is concerned, suffice it to say that many influential theories of lyric poetry celebrate its difference from rational pursuits of knowledge.

By the middle of the twentieth century, the positive valuation of poetry's exemption from rationalism had become so entrenched that poets with widely divergent styles shared this model of poetic value. A year before Williams distinguished poetry from the news, E. E. Cummings offered this prayer to the muses: "may my heart always be open to little / birds who are the secrets of living / whatever they sing is better than to know / and if men should not hear them men are old."[11] He makes a negative judgment of knowledge, claiming that "whatever" birds sing "is better than to know," but birdsong itself discloses "the secrets of living," truths inaudible to the old. Cummings thus maintains the romantic affiliation between poetic inspiration and childlike naiveté. Seven years earlier, in his "Notes Toward a Supreme Fiction," Wallace Stevens likewise deploys a figure of youthful exemption as he advises an ephebe about poetic inspiration: "You must become an ignorant man again / And see the sun again with an ignorant eye / And see it clearly in the idea of it."[12] The imperative to "become . . . ignorant . . . again" does not permanently abrogate knowledge; it allows the would-be poet to see clearly in visual and ideational senses. Just as Cummings's birdsong is "better than to know" but still discloses "secrets," so Stevens's rhetoric about poetic ignorance remains flexible, commutable into knowledge.[13] Such flexibility proves crucial to these rhetorical exchanges, which first reject the strictures of rationalism and then embrace the alternative as a new basis for knowledge. As Donald Rumsfeld's idea of the "known unknown" reminds us, the mere predication of an unknown produces it as an object for epistemic adjudication and thus, in some sense, as an object of knowledge. Such ambivalences about what counts as rational, or as knowledge, abound not only in poetics but in a wide range of disciplines. Cummings, Stevens, and Williams share a commitment to the rhetoric of exemption from rationalism as a language of poetic value, a belief that poetry's privilege lies in its distinctness from rational knowledge. The profound differences among these poets in other matters indicate the extent to which this commitment had, by the mid-twentieth century, become a widespread norm.

Since then, poets and scholars have continued to celebrate the lyric for its exemption from rationalism, often while also comparing poems to machines of one kind or another. Allen Grossman, for example, opens his influential "Summa Lyrica" by describing poetry "as having a function in the same way as a machine has a function."[14] Grossman often distinguishes poetry from rational knowledge, and he offers a rejoinder to Plato's famous verdict: "Wherever the philosopher says *per impossibile* the poet shows the way."[15] The lyric, for Grossman, outstrips the philosopher's brand of knowledge and thereby, he believes, offers deeper insights—precisely the kind of rhetorical exchange I am tracking. The chapters of *The Poem Electric* interpret this rhetoric in the more specific, concrete scenes in which its effects become clearest, but Grossman is far from alone in making more general claims about the lyric's technological dimension and its exemption from rational knowledge.

To tell the story of American poetry since the mid-twentieth century, scholars often describe a series of rebellions against various received lyric norms, including the tradition of celebrating poetry's exemption from rationalism. The didacticism of 1970s protest poetry, the Language poets' bricolage of politically symptomatic public discourse, and the conceptual writers' appropriations of news reports and legal documents—all seem to emphasize poetry's rational, informational dimensions, its capacity to deliver knowledge. Nonetheless, the rhetoric of lyric exemption has proved remarkably durable as a framework of poetic value. Many writers who disavow traditionally lyrical modes nonetheless celebrate poetry's difference from rational knowledge, and they thereby sustain a recognizably lyrical rhetoric of poetic value.[16] Experimental writers increasingly view the lyric as lapsed or outmoded, but quintessentially lyrical ideas continue to inform even the most radical literary innovations, as objects either of disavowal or of recovery. When Craig Dworkin and Kenneth Goldsmith titled their influential anthology of conceptual writing *Against Expression,* they staked the coherence of their movement upon a shared opposition to the lyric norm of personal expression, a norm each piece in the collection ostensibly resists.[17] Such attacks on expressivism can be seen as attacks on the lyric per se, for as Mutlu Blasing notes, "the lyric confirms the *necessity* for an 'I' to intentionalize the linguistic code," for an expressive agent to coordinate the play of signifiers.[18] Poetic experiments with chance and randomness, such as those discussed in chapter 2 of the present volume, thus enable writers to ask how poetry might operate beyond the horizon of lyric expression. Plenty of poets continue to write lyric poems, of course, but the specific matter of the lyric's belatedness, of its having begun an afterlife, powerfully organizes both the experimentalist claim that the genre's time has passed and the traditionalist effort to renovate it. For the innovative writers featured in this study, the lyric becomes an increasingly reified object of disavowal. Writing antilyrical poetry and criticism often means further solidifying the concept of the lyric and acknowledging its continued sway over expectations for poetry. This is the sign of the lyric's afterlives: it is dead, and it cannot be killed.

The analysis of lyric exemption, among other persistent lyric norms, makes it possible to supplement historicist accounts of the genre's emergence with a developmental view of its futures. Influential historicist critics have argued that privileging the lyric effaces the historical variability among and within genres.[19] They claim that finer attention to poetic types like the ode, hymn, and ballad provides a richer image of literary history. By this account, the modern subsumption of all shorter poems under the sign of the lyric suggests that historical effacement is central to the genre's function. This account brings welcome attention to the poorly understood histories of certain poetic genres. It also chimes with broader concerns that the lyrical pursuit of timelessness and universality obfuscates important political histories and social differences. Indeed, many lyrics elevate a moment of expressivity, aspiring for an ahistorical universality, and thereby elide the historically variable contexts from which poems called lyrics emerge. But scholars who therefore hope to displace the category of the lyric must perceive no irony in Fredric Jameson's injunction, "Always historicize!"[20] The discourse of literary genres is not only typological, after all, but also normative, and its normative dimension has greater consequences for the actual production of poems. Poets necessarily write in relation to the genres in which they think their work will be positioned, if only by considering that the kind of thing they are writing is a poem. However fervently scholars contest the lyric's deep history, no one can deny that innumerable poets have written in relation to some preexisting, indeed predominant, idea of the lyric. The lyric's exemption from rationalism plays an especially important role in the genre's continuing development.

Lyric conventions do not simply lay a negative groundwork for experimental writers to reject. They affirmatively enable much of the avant-garde work this study reads. Several lyric norms continue to inform new poetry— not only the rhetoric of lyric exemption but also the embrace of personal expression, affective intensity, subjective interiority, formal unity, and detachment from historical contexts. In sum, many recent poetic experiments rely upon surprisingly conventional precepts, and they often solicit surprisingly conventional interpretive methods. In chapter 2, for instance, a poem that Jackson Mac Low wrote using randomized word-selection procedures might seem an anarchic challenge to the intentionalism that Blasing associates with the lyric, but in fact, Mac Low's chance operations serve to detach the poem from historical contexts, the same impulse scholars have critiqued in the lyric. Moreover, close readers will note that, despite its random generation, Mac Low's poetry gives rise to a distinctly individual, expressive voice. This counterintuitive effect, like others in this book, invites us to rethink the expressive technologies of experimental poetry, to reconsider how poets use electronics to rework the parameters of personal expression. While some avant-garde poetry solidifies the coordinates of lyricism by taking it as an explicit object of disavowal, other experimental texts, such as Mac Low's, revise and extend the very lyrical commitments that many assume they reject.

By tracking the afterlives of lyricism in recent experimental writing, this study seeks to do more than read contemporary verse in terms of the ancient distinction between poetic and rational thought. It also develops an alternative to the narratives of division and disjunction that predominate in scholarship about U.S. poetry since the mid-twentieth century. Binary vocabularies for this division abound (modern/postmodern, cooked/raw, academic/Beat, expressive/conceptual, traditional/avant-garde, formal/ free, closed/open, conservative/experimental), and debates about the split go back to the 1950s at least.[21] Neither terminological debates nor proposals for reconciliation have detracted from the central issue of this division itself.[22] In this book, I want to question the structure of this division by demonstrating that experimental poets and their readers remain more invested in lyric norms than they typically recognize, particularly in the expressive dimension of poetic language and its exemption from rationalism. Of course, many poets both experimental and traditional do not fit these criteria because they see poetry as didactic or informational, a discourse of knowledge. If one wished, one could group together those poets who embrace the rhetoric of lyric exemption and those who do not. To do so would reveal less about the cliques and schools that organize verse cultures and more about poets' responses to the broader, deeper lyric tradition. In a sense, *The Poem Electric* tells half of this story, as it tracks the persistence of lyric norms among experimental writers. A complementary story could be told of how electronics have challenged or enabled more conservative poets.

Recent scholarly debates about the lyric have clarified the key terms of the genre's definition, though largely without contesting them. Critics have predicated "lyric" in numerous ways: lyric reading, lyric form, lyric powers, lyric personhood, lyric theory, lyric shame, lyric orientations, and more.[23] While "lyric exemption" joins this litany, it seeks not to redefine the lyric but to focalize and track familiar lyric norms. Almost all recent studies remain in touch with a definition of the lyric as expressing an individual's thoughts, emotions, and perceptions. This framework of individual expression goes back at least to the romantic poets and offers an alternative to the classical and neoclassical theories of poetry as mimesis.[24] Though often scrutinized, the expressivist definition of lyric poetry remains decisive for a wide variety of scholars. Marjorie Perloff, noted champion of antilyrical experimentation, writes that "lyric is understood to be the expression of a particular subject . . . whose voice provides the cement that keeps individual references and insights together," terms to which few conservative critics would object.[25] Virginia Jackson, the most prominent historicist skeptic about the genre, likewise names "brevity, subjectivity, passion, and sensuality" as "qualities associated with poems called *lyric.*"[26] Rather than seek alternatives to the expressivist definition of lyric poetry, *The Poem Electric* joins other studies that trace the limits of its norms.

Less consensus has developed in recent scholarship about the lyric's relation to technology and knowledge. In her study of poetry and cognitive science, Nikki Skillman reads poets who have "a common faith in mechanistic interpretations of mind that threaten to dissolve the ideal unity of the expressive, lyric 'I.'"[27] She affirms the expressivist definition of the lyric but assumes that mechanistic forms of knowledge, especially technological theories of mind, will disrupt the expressive practices called lyrical.[28] By contrast, writers in *The Poem Electric* embrace electronics and other technologies as means to enliven and rework the expressive norms of the lyric. Other studies of the lyric specifically mention its difference from rational knowledge. Robert von Hallberg describes something akin to lyric exemption when he notes that "what is called lyric is . . . a variety of language use differentiating itself from other discourses" and that such a "differential concept of poetry identifies a range of language beyond the orders of most institutionalized communication," most "rational" discourse.[29] Blasing theorizes the lyric's difference from knowledge still more explicitly. She argues that poetic language poses a "fundamental challenge to any rational discourse in pursuit of truth" because it functions as "a nonrational linguistic system that is logically and genetically prior to its rational deployment."[30] Meanwhile, Jackson describes "our current sense of the lyric" as a "persistent confusion," by which she means to question the lyric's preeminence. But as Jackson recognizes, one can also read this confusion about the lyric as a lyrical confusion, another guise of the lyric's difference from knowledge.[31] As I argue in chapter 1, Jackson's critique operates through tropes of historical loss and interpretive uncertainty that again distance the lyric from knowledge, even as they destabilize the genre's coherence in other ways. *The Poem Electric* hopes to complement these accounts by identifying the lyric's exemption from rationalism as key to the genre's continuing influence. In surprising ways, electronics enable poets and their readers to rework the expressive dimensions of the lyric, including its affective intensity, its spontaneity, and its subjective depth.

THE LIMITS OF "INFORMATION TECHNOLOGY"

The writers central to this study share a key innovation: they bring electronics into the service of a literary genre that is commonly valued because it does *not* offer the kinds of information and rational thinking normally associated with such devices. When poetry's value lies in its exemption from rationalism, we might expect poets not to embrace electronics, since these serve to gather, organize, and transmit information. The deeply logical structure of a tape recorder or personal computer seems incompatible with the lyric poet's search for alternatives to rational knowledge. Indeed, some poets respond to "information technology" by retreating into literary Luddism, framing poetry as the last refuge of some archaic value our technology seems always about to liquidate. An influential analysis of this position appears in C. P. Snow's account of "the two cultures" in his 1959 Rede Lecture.[32] In American poetry, the Fugitives and their inheritors anticipate Snow's formulation by some decades; its legacy plays out among the New Formalists of the 1990s and many mainstream poets today. In the scene of consumption, these Luddite responses fuel the recent vogue for mechanical typewriters and fountain pens, the countless op-eds proclaiming e-books inferior to print, and such oddities as a kit for turning a typewriter into an iPad keyboard and dock. In each case, literary refinement coincides with a preference for more primitive technologies. If poets and their readers view the lyric and the general literary value for which it stands as more or less explicitly opposed to the rationalization and informatization of daily life driven by electronics, then they might naturally defend traditional literary values by rejecting electronics.[33]

Another possible response is to reconfigure the logic of poetic value, embracing poetry that does, after all, deliver information and aid critical thinking. Such a transvaluation of rationalism within the discourse of the lyric enables an affirmative embrace of informational technologies and techniques. The enduring desire to separate poetry from the technocratic world of business documents and schoolbooks makes this response perhaps less popular than the former, but several influential poets and critics of the Western avant-garde have found it attractive.[34] The Language writers and conceptual poets, for instance, often use densely informational texts as source materials. Poets who experiment with software or data-oriented compositional methods tend to emphasize the structural logic of language, thereby soliciting a rhetoric of poetic value built upon information and rational order. *The Poem Electric* focuses on writers who eschew these possible responses to poetry's encounter with electronics. It identifies a lineage of experimentalists who use electronics not to rationalize poetry but to reassert the lyric's exemption from the coercions of rational thinking.

  The Poem Electric seeks to build upon other recent studies that describe poetry's technological dimensions without claiming that electronics subject literature to a strict rationalism. It responds to these accounts by investigating how electronics enable poets and their readers to animate and rework, rather than reject and surpass, familiar lyric norms.[35] Other influential studies of literature and technology approach electronics as "information technologies" that support the collection, logical manipulation, and transmission of data.[36] This perspective accurately reflects the rationalistic, informational purposes of the engineers who design electronics, but it elides the less orderly ways that writers and readers actually use literary equipment. When scholars describe poetry's technological aspect in informational terms, they risk minimizing the lyric's continued influence, especially upon poets who experiment most actively with technology. They instead focus on poets of the second sort described above, who embrace the rational functions of language and reimagine poetic value in terms of information and critical thinking.[37] These critics have not fully attended to the less rationalistic uses poets find for electronics, uses that actually sustain the distinction between lyric poems and the rationalism that electronics supposedly facilitate.

  Likewise, technologically enabled literary scholarship often affirms the rationalizing effects of electronics by emphasizing quantitative data and empirically oriented interpretive methods. The predominant ways of involving computers in literary analysis, such as code studies, software-aided textual analysis, network and system visualization, and data mining, affirm technology's power to render information, often in highly structured or quantitative ways. These efforts, collectively called "digital humanities" (DH) research, certainly open productive new directions for criticism. In the process, however, they tend to ignore electronics' messier, more complex influences upon how people read and write. As some scholars have argued, research of this kind too often relies upon faulty software tools and data whose apodicticity is assumed in advance.[38] DH criticism frequently marginalizes questions about literature that do not lend themselves to quantitative or intensively structured analysis, such as questions about the difficulties of archival preservation or about a writer's compositional equipment.[39] The emphasis upon highly logical, quantitative analysis may also obscure those less rationalistic qualities for which many readers value the lyric.

It may therefore come as little surprise that DH scholarship rarely addresses poetry as such. Notable exceptions include the online archives dedicated to Emily Dickinson, Thomas Gray, Dante Gabriel Rossetti, Walt Whitman, and others.[40] Though valuable, these sites primarily advance editorial and archival techniques rather than criticism. Of the few DH projects that focus on interpreting poetry, most ignore its formal and rhetorical differences from fiction, treating verse as computationally equivalent to prose.[41] The relative inattention to poetry stems partly from the difficulty of finding computational means to parse the formal complexities of verse. But scholars may also recognize formal complexity as one of several ways that poetry eludes the logical, rationalistic modes of thought that DH methods reward. Indeed, DH scholarship has faced severe criticism, from both within and without, for its politically unappealing instrumentalism. As a collective statement by four DH practitioners puts it, "the same neoliberal logic that informs the ongoing destruction of the mainstream humanities has encouraged foundations, corporations, and university administrations to devote new resources to the digital humanities. Indeed it is largely due to the apparently instrumental or utilitarian value of the digital humanities that university administrators, foundation officers, and government agencies are so eager to fund DH projects."[42] Notwithstanding this important metadiscourse on the politics of DH, actual research in this field rarely critiques instrumentalism, empiricism, or rationalism as rubrics of literary and scholarly value. Such critiques have been fundamental to the most celebrated literary criticism of the past several decades. If previous scholars developed these critiques in response to equivalent attitudes in literary works themselves, then DH research risks overlooking the dimensions of literature, and especially of poetry, that have seemed most valuable to readers.

The instrumentalism of DH research also gives a limited view of technology's epistemic complexities. Media theorists have developed a diverse vocabulary to show that technologies do not always function rationally but also complicate and disrupt knowledge.[43] As they describe technology's variable relation to the rational organization of human thought and action, scholars arrive at differing value judgments. Whereas the lyric poets in this study embrace technological derangements of rational thinking, media theorists often take a negative view of technology's power to disrupt knowledge—a corollary to the technological determinist's more intuitive pessimism. Nonetheless, these accounts give a nuanced sense of the epistemic consequences of electronics, making visible technology's less rational functions.

From an engineering standpoint, of course, the design and production of electronics demand rigorously logical, systematic thinking. However, even the foundational works of computer science develop a rich vocabulary for whatever threatens the tidy rationalism of electronics. For instance, Claude Shannon's breakthrough 1948 article, "A Mathematical Theory of Communication," which inaugurated the field of information theory, not only proposes a quantitative unit of information, the "bit," but also defines its rival, "noise," which disrupts communication.[44] Although Shannon seeks mathematical ways to limit interference in communication, his conceptual moves rely upon various terms for information's rivals, such as noise, uncertainty, entropy, equivocation, ambiguity, and error, each of which signifies a different but related kind of disruption. Also, as we shall see in chapter 2, Shannon develops mathematical models of communication by using chance operations quite similar to those experimental poets employ. Alan Turing's equally foundational 1950 essay "Computing Machinery and Intelligence" proposes that a machine can be said to think not if it passes some test of reasoning but if it can deceive humans: the famous Turing test asks whether a computer can answer questions naturally enough to convince the interrogator that a woman is writing the responses.[45] Turing imagines a scene of technologically mediated communication in which the interlocutor's identity is consequentially uncertain, much like the situation Frank O'Hara's telephonic poetry sets up in chapter 3. While Shannon tackles his problem with rigorous logic and mathematics, Turing leaves open the possibility that engineers might build a thinking machine without knowing how their creation works, a surprising limit on the technician's insight given Turing's own accomplishments as a computer engineer.[46] Turing and Shannon share with today's computer scientists a dedication to rational methods of engineering: they pursue

carefully organized, rigorously logical ways to theorize and design electronics. Nonetheless, the counterposition of such technological rationalism with its others (ideas about deception, uncertainty, interference) plays
nontrivial, productive roles in these foundational texts and in computer
science today. In this sense, "information technology" finds its enabling
limits in rhetoric about the others of rational thinking.

Those with a strongly determinist view of contemporary technoscience
overlook both the messiness of the problems technologists solve and the
social dimensions of their labor. Programmers frequently devise ways to
negotiate uncertainty and, more interestingly, to bring disorder into the
service of computation, such as when randomization makes better weather
models or more realistic enemies in video games. The problem space of
contemporary computer science is not rationally organized but full of unruly challenges that seem to defy logic, making them interesting problems.
Further, no matter how carefully a technology is designed and built, it can
later serve unanticipated purposes, becoming more useful. Last but not
least, technologists deal with not only technical but also social problems;
they stretch the limits of the politically, legally, and ethically possible. The
ongoing proliferation of computers does not chase the horizon of scientific
feasibility but that of profitability. In short, it is easy enough to imagine a
tidy dystopia where computers surveil and constrain our every move. But
this fantasy too narrowly delimits the coercive forces people exert over one
another, and it ignores the continual support that human actions give to
cybernetic systems, without which those systems would swiftly collapse.[47]

The Poem Electric joins other recent efforts to understand technology's
effects upon poetry without succumbing to strong technological determinism.[48] Electronics do guide and constrain us in many ways, but the logic
and intensity of their influence is neither uniform nor predictable. Indeed,
the poets in this study demonstrate how easily a poetic experiment can
disrupt apparently strict technological controls. By the same token, it
seems methodologically important to recall that one cannot deduce the
wording of a poem from a poet's technological situation any more than
from her political, economic, or literary-historical situation; nor can such
contexts, on their own, yield satisfying interpretations of her work. For
this study to attribute a necessary effect to a technology would unduly
narrow interpretive possibilities, since different people use the same tool
differently. The proliferation of electronics also does not mean that technology more completely determines our situation. My machines are more

complex than those in an eighteenth-century publishing house, for example, but they do not discipline my body and behavior as strictly. Hence, even as new technologies lead poets to reconsider the epistemic conditions of their art, the increasing complexity and prevalence of electronics does not necessarily mean that technology's effects upon poetic production have become more pronounced.

For these reasons, I want to scrutinize the implicit, unintended ways that electronics alter poetry cultures. Many poets today actively experiment with electronic platforms.[49] This subset of poetry that emphasizes its technological condition does not, however, illuminate the subtler, more varied, and more pervasive influence of electronics on poetics in general. The effects of new technologies vary not only in intensity but also in kind, so lessons gleaned from self-consciously digital poetry will not indicate the effects of less obvious, less consciously chosen technologies such as the word processor, web browser, or smartphone.[50] The laser printer and PDF have changed my writing practice in ways not only less noticeable but also of a completely different kind than the effects most obvious in a digital scholarly edition or a poem written in Flash, for instance. Given the lyric's embrace of uncertainty over decisive rational clarity, the idea that technology might subtly influence the poet behind her back, as it were, corresponds with the genre's broader epistemic disposition. Criticism focused on more conspicuously technological poetry may elide the technological afterlives of lyricism and obscure the broader, more quotidian effects of electronics.

Media theorists and philosophers have long recognized that quotidian technology proves inconsistently available to our attention. Its evasiveness offers an important cue for investigating the technological basis of knowledge production as well as the limits of our possible knowledge about technology. Some theorists address these limits in terms of technology's embeddedness in everyday habits, its role in constructing the fabric of the ordinary.[51] Others note that close attention to a technological device tends to interrupt its function. As Gilbert Simondon puts it, "the use relation" with a technology "is not conducive to the raising of awareness" about it because the "habitual repetition" typical of tool use "erases the awareness of structures and operations with the stereotypy of adapted gestures," or rote routines.[52] The most influential account of this sort is Martin Heidegger's famous example of the hammer: one hammers most effectively by paying no attention to the hammer as such but employing it directly in the task.[53] Direct attention to a tool interrupts its use, and devices function most smoothly when they escape critical scrutiny. Throughout *The Poem Electric*, I mark this fact with recourse to Heidegger's use of the term "equipment" (*Zeug*): "that which one has to do with in one's concernful dealings."[54] As a tool "in-order-to," equipment recedes from attention and becomes lost in the complex of intentional action.[55] Notably, Heidegger gives "equipment for writing" as a prominent example in his discussion of equipment: "ink-stand, pen, ink, paper, blotting pad, table, lamp, furniture, windows, doors, room"—all these, subsumed into the act of writing, recede from attention and become continuous with an indefinitely extensive world of equipment, of materials "in-order-to" that shape what we do and know, but without our direct awareness.[56]

For Heidegger the meaning of "technology" (*technologie*) is more complex. He traces modern technology to its root in the Greek term *techné*, which refers both to technical instrumentality and to practical technique.[57] Heidegger affiliates *techné* with a poetic function of unveiling or making-present: "*Techné* belongs to bringing-forth, to *poiēsis*; it is something poetic."[58] Here *poiēsis* indicates not only actual poetry but also whatever serves this function of bringing-forth. Heidegger invokes the poetic to distinguish the revelatory aspect of *techné* from the more rational, instrumental functions of modern technoscience.[59] Although the present book frequently shows that poetry's technical dimension helps to distinguish it from rational knowledge, I do not further pursue Heidegger's notoriously idiosyncratic account of the relation between poetry and technology.[60] I do, however, invoke his sense of "equipment" to note technology's tendency to evade critical attention. This evasiveness sets the horizon of possible knowledge about the epistemic consequences of electronics for literary production and reading.

In sum, to emphasize the informational, rational character of electronics may hamper the study of poetry's technological dimensions and misapprehend the actual functions of such devices. From a phenomenological perspective, equipment cannot work usefully while remaining the focus of critical attention; only by evading scrutiny do our tools become fully involved in a task. Of course, the engineers who design electronics employ highly logical, densely informational thought processes to do so, and they have explicitly instrumental purposes in mind. Even here, though, the epistemic foundations of computer science involve rhetorical gestures toward the productive limits of such rational thinking. Once a device has been produced, moreover, its actual uses rarely remain completely within the bounds of the designer's intentions. Given the legacy of the lyric's exemption from rationalism, we might expect poets in particular to make inventive uses of writing equipment and to take affirmative interest in the less rational effects of electronics in daily life. Among literary critics, however, the supposedly rational vocation of electronics unduly restricts the possible ways of imagining their interfaces with literature. DH scholars privilege the computer's ability to render literary texts as informational fields for quantitative parsing.[61] Such work too often ignores how writers and casual readers interact with literary equipment. More important, it blinds critics to the ways that literature, and lyric poetry in particular, stakes its value on its opposition to the very instrumentalist, rationalist values that DH scholars embrace. *The Poem Electric* therefore draws upon histories and theories of electronic media to unpack their less knowledge-oriented functions, which poets and readers take as means to distinguish lyricism from rationalism. This project joins other recent efforts to model strategies of technologically oriented close reading that attend to literary equipment without subsuming poetry within a strongly deterministic technoscientific order.[62]

THE RHETORIC OF LYRIC EXEMPTION

So far, the predicates of lyric exemption remain fairly unspecified. From what, exactly, is lyric poetry exempt, and why? What are the particular mechanics of its exemption? Throughout this study, poets and their readers celebrate the lyric for its difference from a varying but relatively coherent set of terms related to rational knowledge. Hence, we might view the lyric as exempt from the obligation to think rationally, from the injunction to critique, or from the positivism of technoscience. As my second epigraph from William Carlos Williams indicates, however, the rhetorical gesture of exemption itself matters more than the specific predicate from which the lyric gets exempted. By invoking "the news," Williams figures a global information system whose particulars matter less than its difference from poetry. Appropriately, he gives even less information about "what is found" in poems: their difference from "the news," qua difference, matters more. The distinction between lyric poetry and rational knowledge functions rhetorically, rather than metaphysically or epistemologically, because *the claim of exemption as such* shapes the production of poetry and the discourse of poetic value.

Rhetoric against knowledge has important functions beyond poetics as well. For instance, when a courtroom witness responds to every question with "I don't recall" or "I don't know," his studied disavowals may convince the jury that he cannot be trusted. Here one person's claims against knowledge help others to know. Such claims function rhetorically when the logic or grammar of a statement is overridden by its power to persuade or assure. If the jury considers the witness unreliable, this is because it does not interpret his answers logically, in terms of their epistemic claims, but rhetorically, in terms of their performative effects. Such rhetoric is not a trick of slippery phrases; rather, it organizes our attitudes about whole categories of statement. *The Poem Electric* explores many situations in which the rhetoric of exemption ultimately sustains new insights that count as knowledge, like the jury's insight about the witness's unreliability. The specific predicates of this rhetoric may be hard to define partly because of its productive flexibility, its availability as a collaborative other of rationality, knowledge, information, and the many other terms from which it exempts the lyric.

Similar rhetorical exchanges play a central role in the construction of knowledge within the Cartesian phenomenological tradition. Although claims against knowledge have deep roots in Socratic skepticism, Descartes puts them to especially influential use at the start of his *Meditations on First Philosophy*. The inaugural decision of his project is not to assert the *cogito* out of the blue but to disavow knowledge: "I was convinced that I must once for all seriously undertake to rid myself of all the opinions which I had formerly accepted, and commence to build anew from the foundation, if I wanted to establish any firm and permanent structure in the sciences."[63] Descartes thus begins by repudiating all supposed knowledge in order to start more surely from scratch. His repudiation must be rhetorical and not logical, since the linguistic conventions, figural conceits, and methodological commitments sustaining the philosopher's discourse itself remain intact. Otherwise, he would fall silent.[64] Like many writers discussed below, he seems to consider the poetic function of his language, its argumentative tropology, to operate outside the closure of rationalism. His own discourse, in other words, exempts itself from the skeptical bracketing with which his project begins. Descartes frames his radical doubt as "a very simple way by which the mind may detach itself from the senses," which may deceive us, but among his most important inheritors are the modern phenomenologists who derive their philosophy from the senses.[65] Edmund Husserl, for instance, adapts Cartesian doubt into the phenomenological *epoché* that brackets all belief in the actuality of the world: "I have thereby chosen to begin in absolute poverty, with an absolute lack of knowledge."[66] Husserl's repudiation too is flexible and rhetorical, not logical and "absolute," since it remains fundamentally open to renegotiation in the service of knowledge.[67] Both philosophers' opening claims against knowledge involve a figural impulse that looks more poetic than philosophic: Descartes offers an architectonic metaphor, positioning doubt as a foundation for the house of knowledge, and Husserl likens the disavowal of knowledge to a vow of poverty. Whatever its figural supports, the rhetorical exchanges between rational knowledge and its others are no mere dalliances of poets. Some of the most influential models of knowledge production begin with such exchanges, with claims against knowledge that later underwrite renewed certainty.

These shared rhetorical habits clarify the nature of the ancient rivalry between the poet's lyricism and the philosopher's rationalism. Poetry threatens rational philosophy not merely by deceiving with false appearances or false knowledge, as Plato complains. More fundamentally, poetry's externality to the enclosure of rationalism indicates the philosopher's own need to construct an irrational outside as a backdrop against which to define rational thought.[68] This circumstance, in which the rational project necessitates the rhetorical production of irrationality as its constitutive other, simultaneously motivates and results from poetry's ejection from the field of rational discourse. The lyric poet, of course, transvalues this gesture of exclusion, reading it as a happy exemption from rationalist constraints.

I call it "exemption" to indicate this openness to negotiation, this ambivalence about the value of the lyric's externality to rationalism. The word comes from the participle of the Latin verb *eximĕre*, meaning "to take out" (*ex*, "out," prefixed to *emĕre*, "to take").[69] To take something out can mean to exclude, remove, or reject it, but it can also mean to select, distinguish, or privilege, as when we take the best grape out of the bunch. Ambivalence about the implied value of the object of exemption, of taking-out, thus goes back to the word's origin. If the second, more affirmative value-implication of "exemption" remains obscure, then consider that the word shares its root with "example." The latter comes from the Latin noun *exemplum*, a sample or specimen, which is a nominalization of *eximĕre*.[70] The lyric poem's exemption, then, refers not only to its difference from rationalism but also to its exemplarity, its power to illustrate in particulars the more general principles of its being. What, then, might the lyric exemplify? Perhaps lyric poems give examples of exemplarity itself: they emphasize the act of selection or taking-out, drawing from the mundane flow of lived experience something privileged precisely because of its exemption, its having been selected. This emphasis upon the poetic power of selection as such motivates a wide variety of experimental poets, who often appropriate language from everyday life, and it helps to explain my abiding efforts to read individual poems as attentively as possible. But where the lyric poem's value hinges upon its difference from rationalism, what it exemplifies is this gesture of rhetorical exchange itself, a gesture that supports both a rubric for poetic value and a system for knowledge production.[71]

The rhetoric of exemption thus reveals the interdependence of rationalism and its others. This interdependence makes it possible to reevaluate the literary critic's disposition vis-à-vis rational knowledge. If the value of many poems lies in their difference from rational thinking, in their power to make facts and logic less stubbornly coercive, then perhaps strictly rational scholarship cannot fully do justice to the poems we study. On the other hand, if a critic seeks a more ambivalent or distanced relation to rational knowledge, how will her criticism operate, and how can it avoid devolving into a vulgar discourse of personal preferences and idiosyncratic notions? Among others, the discourse of "queer epistemology" has already begun to develop much of the conceptual equipment needed to imagine a less rigidly epistemophilic literary criticism.[72] Although this book is not a work of queer theory, writers with nonnormative sexualities appear in each chapter, and the epistemic complexities of their lives open vital spaces for the rhetoric of exemption. Eve Kosofsky Sedgwick famously proposes "reparative reading" as an alternative to the paranoid "hermeneutics of suspicion" and its "infinitely doable and teachable protocols of unveiling."[73] She recognizes that literature's exemption from rational constraints calls for a literary criticism able to deliberate its own relation to knowledge production. Those working in her wake continue to reconsider how literary criticism might serve purposes other than rational knowledge.[74] *The Poem Electric* addresses multiple works of criticism and poetics that respond to poetry's difference from rationalism by seeking less rationalistic modes of literary interpretation. Far from advocating irrationality, my readings finally affirm the powers of rationalism by taking the rhetoric of exemption itself as an object of knowledge. Nonetheless, this book explores why literary critics express ambivalence about rational knowledge and how this ambivalence transforms criticism, altering the value claims that scholars make about poems and about their own work.

These questions about literary criticism's relation to knowledge productively unsettle the distinction between literary and critical texts. If critics no longer distinguish their works from imaginative literature through a strict dedication to rational inquiry, then a space opens for generic hybrids, for poems that interpret literature and for criticism that strikes lyrical notes. The personal computer may encourage this indistinction between reading and writing: through a variety of encoding/decoding processes, the computer reads as it writes and vice versa. Decades before computers became common, however, the mass media had inspired poets to copy and collage various kinds of public discourse, presenting it as "found poetry" or otherwise appropriating it. Such methods position the writer as a reader who transcribes, edits, revises, cites, and rearranges preexisting text—not unlike a literary critic. Many current experimental writers still embrace these practices that unsettle the difference between writing and reading, between the poet's craft and the critic's. Today, writing often looks like the product of readers and reading like the activity of writers. So instead of deploying terms such as influence, creative revision, or collaboration, this study presumes that a poet's literary engagement with other poetry can function as a reading in the same way that literary criticism does: it constitutes a response to a poem, enriching and redirecting the reader's experience of the earlier text. Likewise, criticism that strikes lyrical notes of its own, rather than privileging sober argumentation, can have poetic effects alongside the poems to which it responds. By trying to read freely between poetry and criticism in this way, I hope to put some pressure on the key generic distinction that the rhetoric of lyric exemption both draws and effaces—the idea that literary criticism does and should produce rational knowledge while lyric poetry is valuable because it does not and need not.

MODES OF LYRIC EXEMPTION

Each chapter of *The Poem Electric* specifies a lyrical mode whose attraction lies in its exemption from rationalism. I call them "modes" not only for the term's general sense of "manner" but also for its technological sense as "any of a number of distinct ways of operating a device or system."[75] Each chapter modulates the rhetoric of lyric exemption into a specific context precisely by working it through a technological system for poetry. The first chapter identifies *affect* as such a mode because it supports claims of certainty that remain immune to critique. No one can refute our feelings or provide evidence against them. Many associate affect with tactility, but this chapter shows that new imaging technologies have linked affective reading with the visibility of writing. Whereas this chapter situates affect as a supremely intimate register of experience, the second chapter turns to *chance*, which renders poetry impersonal. Poets have long believed that chance operations detach the poem from the writer's decision-making and from the historical contexts in which she works. Yet such methods themselves have a distinct history, as do the devices enabling them. Attention to these histories indicates that the exemption of chance from rational systems of causality enables it to facilitate the very free agency and expression it supposedly disrupts. Much as the second chapter finds the impersonal in chance, the third chapter turns to the aural technologies of the telephone and tape recorder in order to argue that *anonymity* has guided experiments with poetic address, the social call of the poem. When we cannot identify the disembodied voice on a tape recording or at the other end of a telephone line, the anonymous voice imbues our social world with a sense of estrangement and distance, upsetting our hope for a social space in which everyone has a distinct and legible name. This chapter hears such anonymous contact in the apostrophic calls that resounded through the midcentury mediasphere, arguing that these anticipate current anxieties about online anonymity and network surveillance. The final chapter describes *improvisation* as a method of poetic timing that pursues exemption from the rationalism of advance planning. Improvised performances do not pause to critique a situation and lay a course of action; they open onto the unforeseen event of invention, disrupting the preordained. Recordings of improvisation, whether on tape or MP3, may dampen the spontaneous effect of live performance, making it repeatable and predictable, but the availability of such recordings also broadens the possible audience of poetic improvisation and sets the conditions for its study. The book concludes with a brief reflection on the technological conditions of contemporary experimental writing in order to adduce the function of lyric exemption in more recent poetry. *The Poem Electric* regards these modes of lyric exemption as broadly exemplary without suggesting that they exhaustively catalog lyric poetry's ways of differing from rationalism.

Chapter 1, "Affect: The Possessions of Emily Dickinson," situates affective response as a mode of lyric exemption facilitated by the technologies that make a poem visible. It examines how various imaging equipment, from photo-offset printing to digital archives, has informed Emily Dickinson's reception since the 1980s, increasing the visibility of her manuscripts and eliciting affective interpretation. Studies of the manuscripts recover Dickinson's view that "the look of the words" carries a powerful affective charge. Instead of providing more information, manuscript images offer "no message to decode," as Susan Howe puts it.[76] They unsettle the shapes of poems we thought we knew, occasioning claims of intimate contact with the poet. Such accounts position affect as a collaborative other of knowledge, exempt from critique. In the process, the equipment of affective reading frames Dickinson within a poetics of the visual field that emerged after 1945 and that still shapes experimental writing. Poetic reworkings of Dickinson by Jen Bervin, Janet Holmes, and Michael Magee use Google and graphics software to reimagine her work, unpacking what critic Marta Werner calls "the iconic implications of the manuscript."[77] Here and in recent Dickinson criticism, electronic visualizations help to situate affective interpretation as an alternative to knowledge-oriented reading practices.

Beginning with Susan Howe's influential *My Emily Dickinson* (1985), affective responses to Dickinson deploy a multivalent rhetoric of possession to figure contact with the poet. A "possession of Dickinson" can mean an object she owned (especially a manuscript), a claim of social or sexual privilege, or a sense that her poems render us affectively possessed, captive to powerful feelings. Responding to the rhetoric of possession in theories of affect by Virginia Jackson, Ruth Leys, and others, this chapter argues that the attraction to affect stems from the assumption that it constitutes our most closely possessed register of experience, that feelings are our own-most. By contesting this faith in affect as a solid interpretive foundation, the chapter asks how the possessive rhetoric of feeling has shaped not only Dickinson's reception but also the affective economies of feminist politics that aim to destabilize patriarchal notions of possession. It then culminates by tracing the different versions of "Safe in Their Alabaster Chambers," one of Dickinson's most famous poems, as the text passed first between Emily Dickinson and her beloved sister-in-law Susan, then among newspaper and anthology editors, and finally into the internet archives of Dickinson's writing. By asking who possesses the poem, and when, and in what sense, we see how affective possession enables forms of writing and reading that consider themselves exempt from rationalist constraints.

The second chapter, "Chance: Gertrude Stein, Jackson Mac Low, and *A Million Random Digits*," turns from the supremely intimate register of affect to the impersonality of randomness. It explores Jackson Mac Low's chance-operational rewritings of Gertrude Stein in order to interrogate the common view that chance helps poets escape "the time-linear cause" of historical contexts and leaves poems "'uncontaminated' by the composer's 'ego,'" as Mac Low puts it.[78] To write his *Stein* poems, Mac Low did not use dice or a deck of cards but a strange reference book called *A Million Random Digits*, a book of numbers produced by the RAND Corporation in the 1940s to help scientists at Los Alamos design nuclear weapons. To interpret the *Stein* poems through their equipment illuminates the history of chance as a poetic device and tests its power to detach poetic production from historical contexts and authorial intentions. While many see chance as unsettling contemporary aesthetic and political influences, Mac Low's equipment insinuates a forbiddingly violent history within the scene of writing. It thus suggests that even a poetics intent on decontextualizing itself emerges from specific material histories. Attention to Mac Low's methods clarifies his differences from Stein and indicates how the poetics of chance has developed from Stein's time to our own. Together they disclose a lineage of poets who take chance as a counter to rationalist ideas of historical causation and poetic expression.

A Million Random Digits has found less sinister uses in recent decades, but it still recalls the military projects from which many of today's electronics emerged. It thereby reminds us that the same devices enabling our leisure and play also facilitate state violence. Indeed, many electronics were designed first for military purposes and only later adapted for consumer enjoyment. The so-called RAND book also links Mac Low's work with the unfinished history of efforts to computerize chance. Despite its strictly logical structure, computer software performs many tasks that require an element of randomness—from modeling fluid dynamics to simulating a more lifelike videogame opponent. Such tasks rely upon a mathematics that could not have developed without *A Million Random Digits*. Software's deterministic character prevents the direct generation of random numbers, and today's solutions to this problem rest upon the foundation of the RAND book and the military projects that inspired its creation. In the

Stein series, then, Mac Low does not simply deploy an impersonal, randomized procedure; he explores the turbulent interplay among causation, accident, and decision that chance operations set in motion. The *Stein* poems raise challenging questions about the importance of textual details and authorial identity for the interpretation of randomly generated texts. Mac Low's equipment also opens a sense of technological history focused not on instrumentality and logic but on the less orderly play of chance that electronics continue to support.

The third chapter, "Anonymity: Frank O'Hara Makes Strangers with Friends," situates the poetry of Frank O'Hara alongside the electronics he often discusses. It finds in his poems a telephonic discourse of anonymity that shrouds identity and social exchange in uncertainty. O'Hara is widely remembered for his popularity. In a eulogy for the poet, the artist Larry Rivers declared, "Frank O'Hara was my best friend. There are at least sixty people in New York who thought Frank O'Hara was their best friend."[79] The poems at first seem to support such images of O'Hara as a socialite, since they often call out to specific friends or to an unnamed "you" of his social circle. Read closely, however, O'Hara's techniques of address evoke anonymity, social detachment, and self-estrangement. His most resonant work utilizes apostrophe to figure such isolation: "if there is a / place further from me / I beg you do not go."[80] This chapter draws upon theories of apostrophe as a misdirected structure of address to argue that, for O'Hara, apostrophe marks the absence of a more direct social call, the telephone call. His poetic address highlights midcentury technologies and techniques of sociality even as it delays and disrupts such contact. O'Hara contrasts his poetic address with machines such as the telephone and tape recorder, as when he recalls writing a poem to a lover: "While I was writing it I was realizing that if I wanted to I could use the telephone instead of writing the poem."[81] Many read this famous sentence as equating poems and telephone calls, but O'Hara writes a poem to his lover "instead of" calling him. In O'Hara's rapidly mediatizing era, communication becomes so easy that its technical supports recede from attention. His apostrophic calls suggest that our social techniques come between us even as they connect us. For O'Hara, the technicity of our social devices renders us anonymous, deferring social contact.

Decades before today's debates about the politics of online anonymity, O'Hara recognized that the telephone, television, and other electronics animate a play of identity and anonymity as we communicate through them. In the days before caller ID, for instance, one could never be sure whose voice waited behind the telephone's anonymous ring—an uncertainty several of O'Hara's poems dramatize. For him, such namelessness invites us to think critically about how we call out to others and how we address ourselves. O'Hara distinguishes between such banal equipment as the telephone, whose technical nature gets forgotten as we use it, and poetic techniques that foreground the artificial character of our social calls. The latter indicate that to think critically about the electronics enabling social contact is also to interrupt their daily usefulness, rendering us uncertain of those we hope to reach. Anonymity thus marks the latent technicity of our social calls, and O'Hara's poems play counterpoint to this estranged sociality by reworking the techniques of address.

The final chapter, "Improvisation: Amiri Baraka, Allen Ginsberg, and Spontaneous Politics," reads improvisation as a strategy through which Baraka and Ginsberg develop alternatives to rational planning in performed and written poetry. "Improvisation" stems from a Latin root meaning "unforeseen." Skilled improvisation requires practice and preparation, but at its core, to improvise is to act without knowing in advance what you will do. During the postwar years, Baraka and Ginsberg were central to what Daniel Belgrad calls "the culture of spontaneity," a broad group of artists and intellectuals who viewed improvisation as politically and aesthetically valuable because it liberates us from the planned, the expected, the predetermined.[82] It seems to free artists from the pressures of historical backgrounds and contemporary norms. But improvisation has a politically salient history of its own. As an artistic technique, it developed from a variety of African American aesthetic practices, most notably jazz and other traditionally black music. Baraka (whose birth name was LeRoi Jones) sought to prioritize these African American artistic traditions when, after the 1965 assassination of Malcolm X, he separated from Ginsberg and his other friends in the white avant-garde, moved to Harlem, founded the Black Arts movement, and eventually changed his name. Baraka would later argue that political attachments to spontaneity during the 1960s proved misguided because improvisation displaced more organized and effective forms of political action. His poetry and performance techniques, however, remain highly improvisational from the mid-1960s forward. Through improvisation, Baraka links his poetry with the deep history of African American music, which he and his readers view as a politically vital tradition. Improvisation enables Baraka to dramatize the liveliness of discourse about racial politics in the postwar and contemporary United States. His historically engaged improvisations caution against the unqualified celebration of spontaneity as a means to escape rational planning. Improvisation might liberate artists from such orderly predetermination, but such escapism must not obscure the histories of the African American arts from which improvisation emerged.

Although Ginsberg does not emphasize improvisation's African American origins, his work makes clear how improvisation complicates the effort to read poetry in relation to its equipment. In the same years when Ginsberg and Baraka became estranged, both made improvisation more prominent in their performance styles. By the late 1960s, Ginsberg had become a prominent spokesman for the counterculture, often appearing on television, radio, and other mass media. He also had begun to integrate chanting and song into his performances as a way to make them more spontaneous. In his media appearances, comments about his writing process, and recordings of his poetry performances, Ginsberg explores the tensions between spontaneous self-expression and the technologies that preserve poetry—whether on paper, audio tape, or computer memory. By capturing the precarious moment of unforeseen action, such technologies make it reproducible, enabling us to encounter improvisation outside the context of live performance, but they also dampen the verve of spontaneous invention, making it predictable and rote. It is possible to record an improvised performance as it happens, but will some trace of the artist's spontaneity remain evident in the recording when it is the same every time we return to it? When viewed alongside the political history of artistic improvisation in Baraka's work, Ginsberg's experiments in poetic spontaneity help us to address such questions about improvised poetry's relation to its equipment. As Ginsberg understood, the equipment that delivers poetry to its audience shapes the rhetoric of poetic improvisation as greatly as does the equipment that initially captures a poet's words. This chapter therefore reads the performances of Baraka and Ginsberg in relation not only to the devices used to record them but also to the online archives of literary audio that make recorded poetry readings more accessible.

The Poem Electric concludes with a short reflection, "Lyric and Objecthood," that generalizes the interpretive methods of this study by bringing them into conversation with three recent computer poems, works constructed with animation software and necessarily viewed on a computer. To track the rhetoric of lyric exemption in the era of poetry's electrification, this book combines technologically oriented interpretation with more conventional literary hermeneutics. It seeks to focus on the materiality of poetic texts but also to avoid the kinds of determinism that privilege historicist interpretation and cast lyricism as a mere epiphenomenon of broader technohistorical processes. Responding to Michael Fried's influential essay "Art and Objecthood" (1967) in this conclusion, I address recent computer poetry that carefully deliberates how it figures its own objecthood while also reworking familiar lyric norms for computational environments. The effect of this dual effort, an effect I call "lyric objecthood," underscores the challenges of tracking the material dimensions of computer poetry, especially its physical relation to the human bodies that produce and consume poems. Through brief readings of computer poems by Tan Lin, William Poundstone, and Young-hae Chang Heavy Industries, I explore how the afterlives of lyricism continue to unfold as more poets use software to write and share their work.

1 Affect

The Possessions of Emily Dickinson

Emily Dickinson has occasioned a surprising number of claims to possess her, perhaps more than any other poet in the language. The possessions of Dickinson begin with Susan Howe's influential *My Emily Dickinson* (1985), though Howe borrows the formulation from Dickinson herself. Subsequent critics echo Howe's title: Alicia Ostriker's "Re-playing the Bible: My Emily Dickinson"; Karen Gee's "'My George Eliot' and My Emily Dickinson"; Cristanne Miller's "Whose Dickinson?"; Annie Finch's "My Father Dickinson"; Lori Emerson's "My Digital Dickinson"; and Vivian R. Pollack's *Our Emily Dickinsons*; as well as poetic responses such as Michael Magee's *My Angie Dickinson* and Janet Holmes's *The Ms of My Kin*. By comparison, in the ninety-five years between the first appearance of Dickinson's *Poems* and that of Howe's book, only two commentaries begin their titles with "my," and neither positions Dickinson as the object possessed.[1] Why all the possessive rhetoric in recent years?

This possessive trope emerged as electronics helped Dickinson's readers to elevate affective response as a basis for lyric poetry's exemption from rationalism. New imaging technologies have encouraged responses to Dickinson that hinge upon the visibility of her handwriting, her material possessions, and her face. Although "information technologies" such as the personal computer play a key role in making Dickinson more visible, these images do not provide readers with more information. Rather, visual encounters with what Jack Spicer called the "orthographic chaos" of Dickinson's handwriting disrupt and defer knowledge, unsettling the shapes of poems we thought we knew.[2] Holograph images offer "no message to decode," as Howe puts it.[3] Instead, such images act as sites of affective response. Affective interpretation has become popular in many literary fields, but affective readings of Dickinson exemplify this trend with particular vividness because of the poet's reputation as a sentimentalist and the complex technological lives of her texts.[4] Affect's privilege stems from its difference from rationalism, its apparent immunity to critique: no one can refute my feelings or provide evidence against them. Far from rationalizing our responses to Dickinson, electronic images of the poet and her work encourage affective interpretations that distinguish themselves from informational modes of engagement. In the discourse about these images, affect sustains the rhetoric of lyric exemption, as electronics enable poets and their readers to develop affective alternatives to rational protocols of interpretation.

Poets and scholars interested in affect distinguish it rhetorically from rational thinking. The prevalence of this distinction suggests that affective reading proves attractive because affect appears *exempt from rationalism*. Such claims of exemption are rhetorical in two senses. First, they remain labile in relation to the modes of thought they disavow. Those who privilege affect by distinguishing it from rational thought still view it as capable of underwriting various claims of certainty. Feeling is immune to critical scrutiny but remains available as a basis for claims of knowledge. Through such exchanges, the rhetoric of lyric exemption ultimately serves to reconfigure what counts as knowledge. The production of these specifically poetic kinds of knowledge could not proceed without the rhetorical detachment of certain modes of thought, such as affect, from the enclosure of the rational. Second, as it exceeds the logical signification of a strict grammar, rhetoric refers us to the excessive phenomenality of the encounter with a text. Some view Paul de Man's discussion of rhetoric in this sense as a materialism, but I join de Man in seeing the reductive destination of rhetorical reading as sensorial rather than material. Affective readings of Dickinson appeal not to the material singularity of the poet's manuscripts but to the visual encounter with images of them. The experience of seeing Dickinson's manuscripts and other possessions, through whatever imaging equipment, remains important for many poets and scholars responding to her.

Hence, this chapter is a study not of Dickinson but of her recent reception, an effort to track the habits of interpretation that distinguish affective response from rational thought. As such, it responds to ongoing debates about Dickinson's significance for theories of the lyric genre and about the relation between her manuscripts and the many typeset editions, transcriptions, and facsimiles of them. The chapter ends by discussing possessive responses to Dickinson by two poets working today in order to show that the visual and affective models of reading developed in studies of Dickinson also inform contemporary poetic production.[5] Where some scholars hope to correct the errors of earlier readers by returning Dickinson's work to the rich historical particularity of its time, I argue that critics over the past few decades have constructed an image of Dickinson for *our* time—an image that has shaped ideas about the history of American poetic experimentation, the nature of affective response, and the role of vision in literary interpretation. These revisions of Dickinson stem from technological and poetic developments the poet could not have anticipated, but in her poems and letters, Dickinson in fact lays important groundwork for the possessive model of affective interpretation. She often figures powerful emotion as electrical, anticipating the affective responses today's electronics make possible. Although new technologies have made it much easier to see Dickinson's handwriting or her face, even the first collections of her work in the 1890s were published by early electrical methods; the history of making Dickinson's writing visible, in print and in holograph, has been a history of electronics from the start. By the same token, although most scholars of affect link it with the sense of touch, Dickinson recognizes the visual register as an important conduit of feeling, as it has been in the affective responses to images of her handwriting and her face. Today's devices have equipped some readers to take possession of Dickinson more decisively, rendering an image of the poet that would have seemed unfamiliar to the historical Emily Dickinson as she lived. Still, today's electrified Dickinson does enable her readers to see affect sustaining lyric exemptions from rationality.

FACING DICKINSON

The story of Philip Gura's photograph illustrates how the rhetoric of possession organizes these affective responses to the visible Dickinson. Browsing eBay in April 2000, Gura noticed an ad for a "vintage Emily Dickinson albumen photo." If authentic, it would be only the second known photograph of the poet after childhood. Gura, a professor of American literature, won the item for $481. He recounts finding and attempting to authenticate it in an article, "How I Met and Dated Miss Emily Dickinson."[6] Along with the feminizing "Miss," Gura puns on "Met and Dated" in his title, inviting us to read his story as a romantic liaison—an affair of the heart, as it proved to be. The article privileges Gura's affective experience, rather than anything he or we can learn from it. He boasts not of finding key biographical evidence but of having "experienced what it really meant to possess, and be possessed by, a picture that may show Emily Dickinson at the height of her creative powers." Like many of Dickinson's readers and like the poet herself, Gura understands affective intensity in terms of possession—as a mode of capture that upsets the difference between possessing something and being possessed by it, between the *who* and the *it* of affective experience. He privileges affect by distinguishing it from rational knowledge. For him, "what it really meant" to possess this photograph refers not to any matter of fact but to an affective response exempt from rational scrutiny, not subject to debate or verification.

By emphasizing affective experience, Gura brackets uncertainty about the photograph's authenticity. Affect thus enables a rhetoric of lyric exemption: we do not, cannot, need not know whether the picture really shows Dickinson, since our feelings matter more. As it happens, most now think it is not Dickinson. The definitive analysis, by George Gleason, puts things bluntly: "The image is not of Emily Dickinson."[7] In an addendum to his article that anticipates Gleason's verdict, Gura admits it is "most sensible to assume" the picture is inauthentic, as others do, but he states his own position earlier: "I know it is she, even if I cannot yet absolutely prove it." One almost hears him plead that he can *feel* it. He consoles himself with the thought that, if it proved genuine, he would "have to find a home for it in some institution." Until then, this image and its affective enablements remain his personal possession: "I can look at this image every day and thus perhaps get as 'close' to this elusive woman as anyone can. It's a delightful feeling." Gura at once avows emotional certainty about the picture and disavows any evidentiary logic that would either devalue it as fake or confirm its authenticity, and thus dispossess him. Possession of the photograph helps Gura understand visual contact with Dickinson's image as an avenue of affective contact with the poet herself. His visual pursuit of such contact relies upon affect's exemption from rational thought, its putative immunity to critique. The misogynistic undertones of Gura's piece would make it easy to dismiss if not for the fact that so many others dealing with images of Dickinson and her handwriting deploy similarly possessive rhetoric to frame affective response as attractively distinct from rationalism.

In 2012, Martha Nell Smith announced the appearance of another photograph, but the debut of this picture reveals less about Dickinson herself than it does about the technological systems of visualization through which her image continually emerges. The new daguerreotype, dated to 1859, shows Dickinson seated with another woman, Catherine Scott Turner Anthon. So far, efforts to authenticate it seem promising. After comparing the new daguerreotype with the known picture of the poet, the ophthalmologist Susan Pepin concludes, "I believe strongly that these are the same people."[8] Pepin measured ocular anatomy in the two pictures and observed similar astigmatism in the right eyes. Her report, part of an exhibition in the *Dickinson Electronic Archives,* is accompanied by two "xray-type images of Emily Dickinson's right eye," cropped from the two daguerreotypes (Figure 1). These appear to be color inversions of the originals, rather than x-rays. Their negative shading makes Dickinson's astigmatism easier to discern, but it also lends the exhibition an uncanny mood. Here is the blue, cycloptic gaze of a vaguely alien being, as much a product of contemporary technoscience as of early photography. Dickinson's imperfect vision was already well known, so the signs of astigmatism serve more as metaphotographic evidence, establishing identity between two pictures rather than providing new information about their shared subject.

These techniques of visibility do not emphasize a particular image of Emily Dickinson so much as they make it possible to reflect upon the contemporary photo-ontological technologies that condition the poet's appearance. In his famous essay on mechanical reproducibility, Walter Benjamin identifies "the fleeting expression of a human face," reproduced through the photographic apparatus, as "a last refuge for the cult value" of aura, whose impending decline he celebrates.[9] *The Dickinson Electronic Archives* sustains this affective photo-ontology through an exhibition featuring several videos by Tom Tamm that compare the new daguerreotype with other images of Dickinson by slowly dissolving between them. In the video comparison of the new picture with the familiar 1847 daguerreotype, Tamm adds eight thin guidelines that form a hash across Dickinson's face, marking the boundaries of her eyes and nose (Figure 2). This matrix of photographic identity remains steady as one image fades into another, indicating that it is the same woman because her eyes and nose are the same size. Even as it appeals to the authority of photographic evidence, however, the video animates the slipperier terrain of digital photo-manipulation: given their different sizes, and given the overlaid guidelines, these two faces of Dickinson

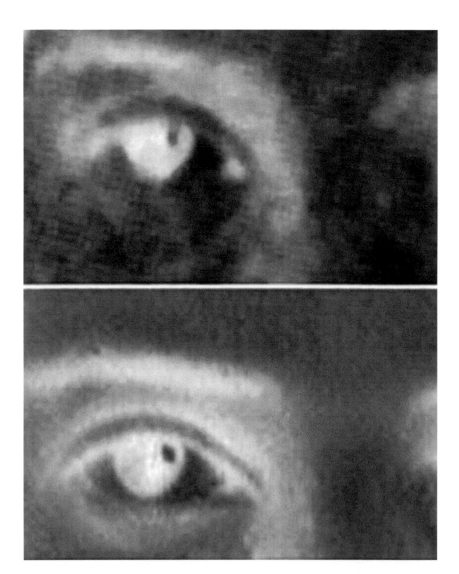

Figure 1. Negative images of Dickinson's right eye, taken from the known daguerreotype (top) and the new one, show an imperfection suggesting astigmatism. "A New Daguerreotype of Emily Dickinson?" *Dickinson Electronic Archives.* Reprinted by permission of the Dickinson Editing Collective.

must have been manipulated in order to make them fit together so neatly. I do not mean to question the investigators' methods or dispute the new picture's authenticity. Rather, the circulation and manipulation of these Dickinson pictures, whether they are genuine or not, indicate that the visual rendering of a poet and her work constructs the historical actuality to which the pictures refer. The photographic exhibition of her face as a site of identity and affective intensity serves, as Benjamin indicates, to sustain a certain cult value around the personhood of Emily Dickinson. Such constructions owe a great deal to the contingent development of imaging technologies and their attendant rhetorics of photographic actuality. As I will show, these technologies do not serve primarily analytic or informational ends but foster a rhetoric of lyric exemption that empowers affective reading practices to reshape Dickinson's legacy.

In her essay on the new daguerreotype, Smith suggests two ways it might remake our image of Dickinson. First, because it shows her not alone but "with her arm around another woman," this picture helps dispel the idea

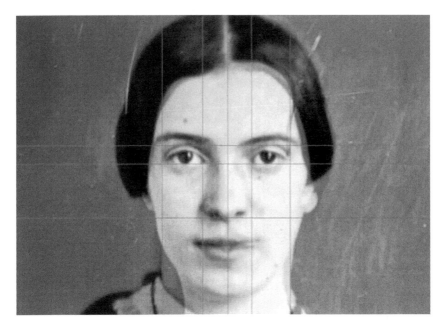

Figure 2. In a video comparing the two daguerreotypes, overlaid grid lines remain stable as one image of Dickinson's face fades into the other. "Daguerreotype Comparison," *Dickinson Electronic Archives.*

of Dickinson as solitary and reclusive.[10] The fact that she appears with a woman further strengthens the case that she preferred the social and perhaps sexual company of women. This seems right, but since the 1980s, Smith and other influential critics have not needed this picture in order to offer persuasive accounts of Dickinson's lively (homo)social engagement. Second, Smith notes that Dickinson was only sixteen years old in the pre-existing 1847 daguerreotype, at which time she "had not written . . . the powerful poems we know so well."[11] In place of this dour adolescent—an image her siblings considered "too solemn" and hoped to suppress—the new image shows "a bold, assertive woman in her late twenties."[12] This Dickinson appears adult, nearer to the "height of her creative powers," as Gura had it. There are obvious advantages in seeing Dickinson as a grown woman rather than a doe-eyed ingénue, but fetishistic habits are not easily overcome.

For me, the new daguerreotype's most striking feature is not Dickinson's age or her companion but her facial expression. While she looks rather gloomy in the earlier picture, here the left side of her mouth curves subtly up, verging upon a smile without fully arriving there. If one can call it a smile, it is an archaic smile—not because it reflects spiritual wellbeing but because its occasion remains uncertain, lost to history. Is this an expression of contentment? A smirk? Is she really smiling at all? The picture challenges affect theorists' common belief that "our facial displays are authentic read-outs of discrete internal states that constitute our basic emotions."[13] It instead resonates with Michael Snediker's claim that the Dickinsonian smile advances a queer, inchoate optimism. For him, Dickinson's smiles "argue against their own legibility or interpretability" because a smile in her poetry refers "less to any emotional circumstance before or beyond it than to the textual event of the poem, itself."[14] Snediker identifies her smiles as a special case that he uncouples from a broader economy of masochism and negative feeling that governs affective readings of her work and, as he argues, affective reading more generally. In Dickinson criticism and in her poetry, negative affect casts the face as a visual surface across which legible emotions play out:

> I like a look of Agony,
> Because I know it's true –
> Men do not sham Convulsion,
> Nor simulate, a Throe –[15]

Dickinson finds something to "like" and thus follows the "felicitous persuasions" Snediker seeks to recover, but what she likes is the face's reli-
ability as a barometer of negative affect. Suffering, the poem continues,
is as "Impossible to feign" as "Death" itself. Notwithstanding the archaic
smiles that "elude the eyes" (F 335), Dickinson's apodictic "look of agony"
anticipates the views of Silvan Tomkins and Paul Ekman, two influential
affect theorists for whom the face serves as a reliable barometer of affect.[16]
Dickinson also shares this regard for the face with her contemporary
Charles Darwin, whose *The Expression of Emotions in Man and Animals*
(1872) focuses on facial expressions and was excerpted in the *Springfield
Republican*, which the Dickinsons read.[17]

Two factors mitigate Snediker's otherwise felicitous effort to detach
Dickinson's ambiguous smile from the facial conventions of affective reading. First, to contest the facial figuration of negative affect is to set aside
an interpretive strategy that has pervasively shaped Dickinson's reception
for the past few decades. As Virginia Jackson notes, the negative affects
classifiable as *pathos* have involved Dickinson's manuscripts in a cycle of
loss and recovery by which readers situate them as lyric poems and privilege affective response, bracketing the historical particularity of her texts.[18]
I follow Jackson in acknowledging the influence of such affective reading
strategies upon Dickinson's reception, but I question whether her critique
can interrupt these cycles of melancholic attachment. The present chapter
unpacks the technological background for the affective approaches to
Dickinson in order to suggest consequences broader than the "lyricizing"
effects Jackson laments.

The second factor limiting Snediker's critique of affect's facial legibility
is that Dickinson herself posits this same metonymic equivalence between
the face as a visual expression of affect and the text as a visible surface inviting affective interpretation. Her writings do not simply thematize the
facial expression of affect, as in the verse above; they perform the visibility
of text as soliciting affective response. This slide from the face to the text
as the locus of affective reading appears in the passage from which Howe
borrows the possessive device of her title, *My Emily Dickinson*. Howe cites
an excerpt from Dickinson's letter to her Norcross cousins in which she
recalls reading news of the death of George Eliot:

> The look of the words as they lay in the print I shall never forget. Not their
> face in the casket could have had the eternity to me. Now, *my* George Eliot.

The gift of belief which her greatness denied her, I trust she receives in the childhood of the kingdom of heaven.[19]

Dickinson here presages today's possessive rhetoric and the linkage of affective response with a facial visibility. She writes of belief as a gift one receives (or not), and by attributing Eliot's atheism to her "genius," she tempers her own spiritual "trust" with a suggestion that her faith indicates a lesser creative "gift."[20] Meanwhile, her affectively intense experience of reading unfolds through a visual encounter with "the look of the words." These words have a "look" in the sense of an appearance "in the print" but also in the sense of a personified gaze. The slippage between facing words about loss and facing a lost person anticipates the "confusion between the pathos of a subject and the pathos of transmission" that Jackson traces in lyric readings of Dickinson, analogous versions of the traffic between the *it* and the *who* of possession.[21] For Jackson, constructing a lost textual object in need of recovery gives lyric readers a sense of Dickinson's subjective depth, and for Dickinson, the "look" of the words announcing Eliot's death bears the "eternity" of loss more powerfully than would a face in a casket. This personifying rhetoric of textuality has redounded through Dickinson studies as a "conflation of Dickinson's physical and textual bodies"[22] that equates her person with "an inviting and vulnerable textual *corpus* upon which power can be, has been, and is now being . . . deployed."[23] Despite its disciplinary connotations, the treatment of texts as visible bodies provides both Dickinson and her readers with a welcome sense of affective contact. In Dickinson's letter, the transition from words to a person occurs through the phrase "their face in the casket." We can read "their face" as that of the words she describes looking up from the page. But the phrase also indicates Eliot's face: "their" can function as a singular epicene pronoun, a nod to the open secret of Eliot's gender. Through this ambivalent pronoun, "the eternity" refers both to the permanent loss the words announce and to the words' own imperviousness to time, their fixity "in the print." Eternity stands both on the side of Eliot's heaven-bound soul and on that of the printed words that will, like Eliot's writing, live on for us. Just as Auden says upon Yeats's death, "he became his admirers,"[24] so Dickinson says, upon Eliot's death, she is "Now, *my* George Eliot," no longer her own. Already in the letter that inspires Howe's title, Dickinson affiliates possession with an affectively intense readership and with the visibility of words on the page, the same commitments that shape recent claims to possess her. If she in turn is now Howe's Emily Dickinson, or mine, or yours, then such possessions emerge through an affective reading practice that unworks the distinction between seeing a face and seeing a text, between the *who* possessed by feelings and the *it* that transmits affect to us.

The possessive rhetoric about Dickinson thus emerges not as an object of censure but as a source of analytic traction, improving our grasp of the politics of affective response. On one hand, the Gura affair illustrates how claims of possession underwrite patriarchal efforts to own and control women. On the other, feminist scholars also deploy possessive rhetoric to good effect, indicating its availability for appropriation and reclamation. The political ambivalence and flexibility of this possessive rhetoric enables a web of connotations to develop around it. When writers describe manuscripts and other objects as Dickinson's possessions, as things she owned, they involve her in a discourse of commodification and capitalist exchange, in which the value of an object consists in its interchangeability with the poet herself, a kind of exchange value. The most overtly misogynistic treatments of Dickinson leverage this commodity fetishism toward the further objectification of the poet's body, such as in a condemnable poem by Billy Collins, "Taking Off Emily Dickinson's Clothes," in which the whalebone and mother-of-pearl fasteners of Dickinson's undergarments frame her flesh itself as something that men might not only expose and caress but also buy and sell.[25] Meanwhile, the rhetoric of possession also has spiritual connotations, as in Susan Stewart's theory of "lyric possession," which she describes as a "haunting" ventriloquism wherein the poet speaks with another's voice, as if possessed.[26] Whereas Stewart emphasizes poetry's aural register, the present chapter focuses on its visual, but I follow her in "taking the notion of lyric possession as a description rather than a problem to be overcome or refuted."[27] Each of these guises for possessive rhetoric sustains an ambivalence between objective and subjective frameworks for affective response. On one hand, feeling marks the site of subjectivity per se; affect stems from the very kernel of the subject and, on some accounts, even precedes and facilitates the emergence of subjectivity.[28] On the other, feelings often take hold of us, possess us, in ways we do not choose, and thus affect also renders us as objects, worked over by the world into which we are thrown. In other words, we have feelings, but also they have us. The possessive rhetoric of affect sustains this switching between object and subject, between the *it* and the *who* of affective experience. By articulating the web of connotations glossed above, this possessive rhetoric also lends gravity to the discourse of affective response, highlighting its political stakes.

The remainder of this chapter fleshes out the linkages among imaging equipment, affective reading, and lyric exemption sketched above. Dickinson's publishers have been reproducing her handwriting since the 1890s, but in recent decades, new imaging equipment has made it dramatically easier to see her manuscripts. In this period, critics have become more interested in her material possessions—her fascicles, fragments, letters, enclosures, inks, papers, binding equipment, clothes, furniture, pressed flowers, and other artifacts. To see these materials ratifies the historical actuality of Dickinson's work by projecting it through new technologies of historical transport. The visibility of Dickinson's possessions "keeps recalling modern readers to an archaic moment of hand-written composition and personal encounter," a recall made possible by the very technologies this archaism seeks to forerun.[29] Through a correlative process, readers ignore the enframing functions of electronic imaging devices in order to privilege affective responses to the artifacts these devices make visible. Today's imaging equipment makes it possible to view many of Dickinson's manuscripts on any device with an internet connection, but these technologies do not simply proliferate textual information as so much binary code. At a physical level, of course, our electronics do function according to strictly logical, densely informational protocols. As these machines enable us to look at Dickinson and her possessions, however, we use them in ways that not only elude but also directly oppose such informational functions. Electronic imaging equipment has shaped reading habits that prove far less orderly and rational than the epithet "information technology" suggests. It provides avenues for affective responses that understand themselves as consequentially exempt from rational knowledge.

SUSAN HOWE'S EMILY DICKINSON

Holograph images of Dickinson's writing were published as early as 1891, as a frontispiece to *Poems: Second Series,* but seeing the poet's manuscripts and related possessions has become especially central to her reception in the past few decades. This trend begins with the 1981 appearance of R. W. Franklin's facsimile edition of the fascicles, *The Manuscript Books of Emily Dickinson,* an expensive two-volume affair that has found its way into most research libraries, though relatively few private hands. Drawing upon

Franklin's earlier work with the manuscripts, this collection not only contested Johnson's 1955 ordering of the poems but also, more important, enabled readers to see many of Dickinson's holographs without traveling to Amherst or Cambridge.[30] In the years that followed, further improvements in electronic printing technologies increasingly meant that scholars could expect to include holograph images in their own articles and monographs. This trend has been slowed, however, by the control Harvard University exerts over images of manuscripts in its possession, for which it controls permissions and royalties.[31] By the turn of the millennium, online archives had begun to showcase holograph images, often in color, but until recently, these digital collections either charged access fees to help defray the cost of permissions or included a very limited number of images.[32] Not until 2013 did Harvard and Amherst, in collaboration with other institutions owning Dickinson papers, launch the *Emily Dickinson Archive,* a searchable, open-access, nearly comprehensive digital archive of manuscript images of the poems.

Readers have no doubt learned a great deal from these increasingly accessible holograph images, but if the proliferation of these images has changed how we read Dickinson, it has not done so by fostering more thickly informational reading habits. Indeed, Sharon Cameron's *Choosing Not Choosing* (1992), perhaps the most important study of the fascicles to appear since Franklin's *Manuscript Books,* argues that the handwritten fascicles' many variant wordings and other structural complexities evince Dickinson's preference for "not choosing," for avoiding the kinds of selection and ordering conventional publication demands. For Cameron, in other words, reading the fascicles in their handwritten form does not provide steadier knowledge about Dickinson's poems so much as it disrupts a textual field onto which Johnson and other editors had sought to impose an order her writings resist.[33] As Alexandra Socarides notes in a more recent study of Dickinson's poetic equipment, "the more time I spent with Dickinson's materials, the more they challenged my understanding of what Dickinson was doing and the identity of what she was writing." By praising Dickinson as a "poet who has heretofore eluded, and perhaps will always elude, our grasp," Socarides measures the value of the poems in terms of their power to resist the understanding.[34] Instead of giving more information, the increased emphasis upon what Marta Werner calls "the iconic implications of the manuscript" has helped scholars position affective response as an alternative to knowledge-oriented reading practices.[35] Through the visual encounter with Dickinson's handwriting, affective readings take the place of critical interpretation.

Howe's influential *My Emily Dickinson* brought to prominence the possessive rhetoric that links affective reading with the visibility of Dickinson's handwriting. Howe argues that the image of Dickinson as socially and intellectually cloistered has unjustly obscured her literary and philosophical influences, her engagement with the wide world, and her significance for intellectual history. "Who is this Spider-Artist?" she asks in response to Sandra Gilbert and Susan Gubar's comparison of Dickinson with a reclusive spider, weaving her webs of verse. "Not *my* Emily Dickinson. This is poetry not life, and certainly not sewing."[36] Throughout the book, Howe's possessive rhetoric is motivated by her attention to the affective impact of Dickinson's writing. For her, affective intensities productively displace rote knowledge and then foster new modes of understanding, a process like defamiliarization. She believes Dickinson's first readers had similar experiences:

> The recipient of a letter, or combination of letter and poem from Emily Dickinson, was forced . . . through shock and through subtraction of the ordinary, to a new way of perceiving. Subject and object were fused at that moment, into the immediate *feeling* of understanding.[37]

Intense feeling here occasions a departure from the familiar and an approach to new insight along sensorial avenues, especially, for Howe, visual avenues. The process culminates in a happy conflation of subject and object, of the *who* and the *it* of affective possession. The reader feels herself at once a subject possessing emotions and an object captured by those feelings. Through this conflation, the rhetoric of possession refers both to the artifacts of Dickinson's life and to the affective intensities of her writing. Since Howe's book appeared, the rhetoric of possession has helped make Dickinson central to ongoing debates about the lyric genre, about the role of original manuscripts and holograph images in textual studies, and about the significance of nontextual artifacts from a poet's workshop. As Howe's objection to Gilbert and Gubar's model of reclusive feminism anticipates, by inserting a wobble between subjective and objective senses of possession, between the *who* and the *it* of the possessive copula, this rhetoric has also informed the social theories through which Dickinson's identity-political significance continues to emerge.

Howe revisits Dickinson in "These Flames and Generosities of the Heart," an essay published in 1991 and reprinted in *The Birth-mark* (1993). Both this essay and *My Emily Dickinson* are central to the body of criticism that privileges visual contact with Dickinson's manuscripts, but neither claims that such contact solidifies our knowledge about Dickinson's poems. Responding to *The Manuscript Books*, Howe praises Franklin's challenge to the ordering Johnson had established in 1955. She writes that Franklin has "made available to readers Dickinson's particular intentions for the order the poems were to be read in."[38] Cameron's and subsequent studies of the fascicles suggest that they resist settling into a legible "order," but one enduring claim of Howe's book is that we come closer to "Dickinson's particular intentions" through holograph images than we could through even the best typeset edition. Indeed, as early as 1956, Spicer's review of the Johnson variorum argued that "the printed text cannot be used as a test of her wishes," though Spicer cannot have anticipated Franklin's facsimile edition and does not mention the earlier holograph publications.[39] Howe tempers her praise of *The Manuscript Books* by lamenting not only that it "is huge" and "extremely expensive" but also that "Dickinson's handwriting is often difficult to decipher."[40] How can the difficulty of the poet's handwriting count as a flaw in a book designed to make that very handwriting visible in all its complexity? For Howe, seeing the holographs "is necessary for a clearer understanding of her writing process," even if the reader cannot always make out the words as Dickinson has written them.[41] Instead of providing a basis for textual knowledge, the manuscripts reveal a "nineteenth century American penchant for linguistic decreation" that is lost when the poems are reified in type.[42] For Howe and many others, seeing Dickinson's handwriting shows not only that her work is far less orderly and stable than the typeset editions imply but also that this linguistic destabilization was central to her poetics. The insight we gain from Dickinson's holograph images, then, is that her texts are *less* available to a knowing critical gaze than we might like to suppose.

Howe's emphasis upon the visual dimension of the manuscripts enables her to make claims on their behalf by distinguishing them from typeset texts. In the later essay, she argues that "these manuscripts should be understood as visual productions" negotiating the singular space of the stationery, not as precursors to a typeset poem.[43] She rejects Franklin's claim that the "facsimiles are not to be considered as artistic structures" but as working transcriptions of poems Dickinson often copied or revised elsewhere.[44]

"How can this meticulous editor," she asks, "repress the physical immediacy of these spiritual improvisations he has brought to light?"[45] Instead of clarified knowledge about Dickinson's poems, for Howe, the holograph images provide visual impressions of the "spiritual improvisations," the affective play, that Dickinson's hand originally traced. As the invocation of "physical immediacy" suggests, Howe vacillates between privileging the visual encounter with holograph images, such as those in *The Manuscript Books*, and claiming that these refer us to the more basic material reality of the originals. These two sites of emphasis, visual and material, seem closely linked but in fact have different implications. To stress vision enables a phenomenological mode of interpretation that locates sight and other perception, along with the affective responses these occasion, beyond the grasp of rational deliberation. The phenomenologist brackets sensation as unreliable, true only qua appearance, but the linkage between vision and immediate affective response appears outside this bracketing that sets the bounds of knowledge.

By contrast, Walter Benn Michaels asserts in his critique of Howe that emphasizing the material actuality of Dickinson's manuscripts leads to an impasse. Like Howe, Michaels conflates the visibility of holograph images with the materiality of the manuscripts. He considers the interest in seeing holograph images to be motivated by a desire to ratify the physical singularity of the originals. Howe privileges the visible shapes of Dickinson's marks on the page over the texts they spell, while Michaels claims that "the point of the redescription of the letter as mark is to make it matter" in both senses: to make it material and to make that material dimension important. He argues that, in "becoming 'drawing,'" a Dickinson manuscript "is disconnected from its meaning" because it is no longer an editable, legible text.[46] But surely drawings too have meanings, especially drawings with words in them, and surely these meanings are transferrable through facsimiles, even if they cannot be typeset or edited. As I have begun to show, when Howe argues that seeing the "indecipherable variation" of Dickinson's handwriting brings us to "the other of meaning,"[47] she does not consider the experience "meaningless," as Michaels claims, but positions affective response as an alternative to textualist, informationalist strategies of deciphering.[48] Through *reductio ad absurdum*, Michaels indicts the materialism he sees as the basis of Howe's approach: when Dickinson's artwork is not a text but a physical artifact, every last detail "matters," so this materialism demands fidelity even to the minutest and most accidental features of the manuscript. Everything there is potentially meaningful, so nothing is. However, in both Howe's reading and Michaels's reply, a more careful distinction between visual and material forms of contact would dispel this absurdity. Michaels acknowledges that, as drawings, Dickinson's manuscripts would simply demand a different mode of reproduction, one focused on "getting the right shapes in the right places" instead of "getting the right words in the right order."[49] By this light Michaels simply disagrees with Howe (as Franklin does) about which aspects of Dickinson's work are most important. He links Howe's interest in the "mark" over the "text" with de Man's advocacy for an experiential rather than an interpretive mode of reading, one based in "pure affect rather than cognition," as de Man has it.[50] Though Michaels attacks de Man along with Howe, de Man's deconstruction is based on phenomenological rather than materialist precepts, so it makes sense that an emphasis on bare sensory appearance rather than material substance defuses Michaels's critique. In any case, emphasizing vision makes it easier to track, first, the discourse of visibility Howe's account helped to inaugurate, second, the effects of new imaging technologies as these have shaped the kinds of reading Howe models, and third, the influence upon Howe's reading of the poetic scene from which she wrote.

Before we address Howe's poetic milieu as an influence, another visually oriented presentation of Dickinson's work will make clear that the opposition between text and drawing threatens to obscure a different question, that of formal selection. The recent edition by Marta Werner and Jen Bervin, *The Gorgeous Nothings: Emily Dickinson's Envelope Poems,* presents beautiful color photographs of Dickinson's late poems, most written on envelopes and oddly shaped scraps.[51] Opposite each actual-size holograph, the editors provide strange diagrammatic transcriptions that are visually striking in their own right. Each transcription consists of a digital line drawing that silhouettes the manuscript page and traces any folds or seams. Within this diagram, words transcribed in a sans-serif typeface are situated to match their position in the original, even when this means inverting words, slanting them, or inserting space between letters. Cancellations and stray marks are also reproduced, though the embossing, stamps, and other marks not made by Dickinson are ignored. Given the attentiveness of these schematic transcriptions, it is surprising that the dashes do not vary in length, elevation, or slope. *The Gorgeous Nothings* opens two incompatible perspectives on the visual aspect of the manuscripts. On one hand, the stunning photographs bespeak a fetishistic attachment to the singular appearance of the manuscript page. On the other, the resourceful transcriptions suggest that holograph images can or should be supplemented by a more legible reproduction that retains some visual elements of the original, such as page shape and word placement, but not others. This is a formal selection not different in kind from conventional editions, which simply take a more restrictive view as to the attributes a typeset rendition should reflect. The transcriptions by Bervin and Werner take seriously Howe's idea that the manuscripts are "artistic structures." Treating them like mere texts, as Howe believes most editors do, misses something important about their appearance. But Howe does not call the manuscripts "images" or "drawings," as Michaels implies. If they are structures, their forms and relations can be rearticulated in other media, as these innovative transcriptions attempt to do.

Viewing Dickinson's manuscripts as "artistic structures" thus does not demand strict adherence to their materiality, and perhaps not to their appearance either. Yet many readers resist even the most careful efforts to transcribe these "structures" in type; any perceived departure from the manuscript warrants objection. This resistance stems from an aversion to viewing Dickinson's work as textual, as verbal rather than graphical, and therefore reproducible in type. If these are verbal texts rather than visual fields, then the full range of transcription strategies, from the most conservative editions to the most inventive schematics, varies merely by degrees and not by kinds. To admit as much is to admit that Dickinson's work can be adequately reproduced by *some* means other than visual facsimile, and all that remains is to decide how such transcription should proceed. Readers who resist this conclusion reject the whole range of publishing technologies that operate by means other than visual facsimile. The transcription strategies they reject are fundamentally informational, since these address the manuscripts with a formal attention that extracts a "structure" to be rearticulated in type. For many readers, something ineffable gets left out when a page of Dickinson's handwriting is subjected to this technological reduction. What is essential to Dickinson's poetry—and perhaps to poetry in general—is thus posited as noninformational, beyond the capture of any technology that structures writing in an informational system. The rhetoric against textual structure is a rhetoric of exemption because it leaves unspecified the *je-ne-se-quoi* that is available only in manuscript images. This surplus over any informational rendition of the text is, as we have seen, often posited as an affective surplus. The irreducible complexity and near-illegibility of the manuscript images provides a basis for affective responses that pit themselves against textualist approaches.

Howe's own poetry is best known for its engagement with the page as a visual space and with text as a material substance. Her poems often look more like collage than verse. In fact such techniques, fairly typical for an experimental American poet who began publishing in the 1970s, also inform her writings on Dickinson. Toward the end of *My Emily Dickinson,* a strange textual diagram of Howe's devising traces the "process of Metamorphosis" in the final stanza of "My Life had stood – a Loaded Gun –."[52] This diagram lends her analysis a graphical dimension it would lack if she made equivalent observations in linear prose. Setting aside the diagram's particular claims about a particular poem, it enables Howe to advance her view of Dickinson as a fundamentally visual poet, one whose work solicits visual responses. In "These Flames and Generosities of the Heart," meanwhile, the holograph images are more detailed than in most criticism, often showing just one word. The essay also includes strangely hybrid renditions of the poems. Howe sets the words in type but does not regularize line breaks, and she also reproduces the relative length, shape, thickness, and angles of the dashes and errant pen strokes (Figure 3).[53] It is as though Howe typed up the poems without any dashes and then, consulting the manuscripts, painstakingly drew in the dashes and other marks whose significance seemed

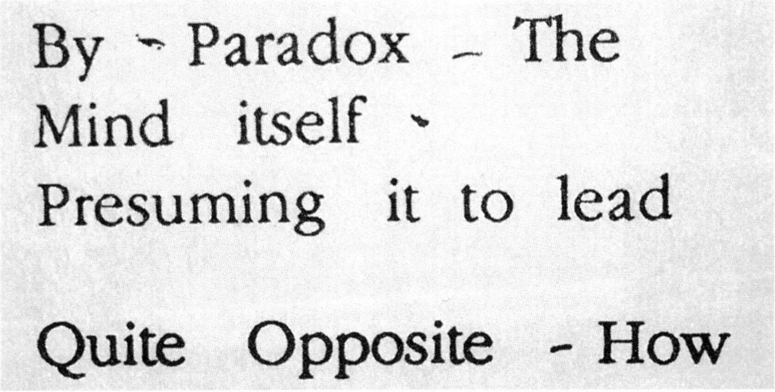

Figure 3. Susan Howe's hybrid transcription of Dickinson's poetry features typeset words with hand-drawn dashes. Photograph by the author.

visual but not textual. Like *The Gorgeous Nothings*, this hybrid approach plays into Michaels's hand. It tempers the commitment to visual impressions by sustaining a desire for interpretive engagement with a legible, printable text.

By intensifying uncertainties about the visual and textual strategies by which Dickinson's work might be presented, these hybrid renditions enable Howe's commentary to accomplish something similar to what she claims the poems themselves pursue: "... [a] deflagration of what was there to say. No message to decode or finally decide ... they checkmate inscription to become what a reader offers them."[54] The visual play of Howe's Dickinson reveals the limitations of reading strategies that bracket the reader's affective response and privilege the construction of knowledge from textual hermeneutics. Howe's readings approach the condition of poetry they describe, unfolding visual textures that frustrate readers' desire to know. The later essay includes a series of one-sentence paragraphs, a set of competing axioms:

> Emily Dickinson almost never titled a poem.
> She titled poems several times.
> She drew an ink slash at the end of a poem.
> Sometimes she didn't.[55]

These propositions juxtapose habits with their exceptions. All true yet glaringly contradictory, together they do not provide information but tease up the bewilderment any reader must feel upon trying to tackle a body of writing so complex as Dickinson's. These sentences also visually insinuate themselves as verse, each indented and isolated on its own line, but they do not violate the formal conventions of the surrounding prose. They thus appropriate a generic conundrum endemic to Dickinson studies. They occupy the same structurally ambiguous space in which Dickinson's holographic "poems will be called letters and letters will be called poems."[56] According to Howe, the editors often elide such generic uncertainty and its implications. She performs the same uncertainty in her commentary, not to resolve it but to insist upon uncertainty itself as a conditioning factor in the encounter with Dickinson. Seeing the poet's handwriting offers something the typeset versions cannot, but this experience sharpens uncertainties in ways that Howe claims Dickinson sought to do. Through the visual play of her commentary, Howe not only describes but also performs some of the "deflagrations" of this visual encounter with Dickinson.

Some recent studies emphasize the manuscripts for historiographic purposes, but scholars responding to holograph images in the 1980s and 1990s offered more proleptic rhetoric. For them, seeing Dickinson's handwriting did not merely support historical transport to her era but made it possible to bring Dickinson into the twentieth century. Jerome McGann positions Dickinson as a modernist born too soon, resisting "the textual conditions of the age of print in which she lived." He claims that she eschewed conventional publication because the print culture of her day would have regularized her formal innovations but that, just a few decades after her death, the new modernist presses would have welcomed her "experiments with a certain kind of what used to be called 'free verse,'" experiments visible in the facsimiles but not the typeset editions. Starting around 1861, he argues, "Dickinson decided to use her textpage as a scene for dramatic interplays between a poetics of the eye and a poetics of the ear."[57] This understanding of the page as both a visual artifact and an annotation for voicing was common among experimental poets in the postwar and subsequent decades. Already in his 1956 review of the Johnson edition, Spicer claims the visual "slant" of punctuation enables him to distinguish dashes from other marks more reliably than Johnson has done, but he interprets the dashes aurally, "as a sign of stress and tempo stronger than a comma and weaker than a period."[58] Such accounts seem to privilege sound over sight, but the text's visibility supplements its aurality. Seeing a manuscript provides the indispensable basis for these readings.

For Howe and many others, this visual supplement to the sound of poetry was most influentially formulated in Charles Olson's 1950 manifesto "Projective Verse." Olson identifies not handwriting but "the typewriter" as enabling the kind of verse he advocates: "Due to its rigidity and its space precisions, it can, for a poet, indicate exactly the breath, the pauses, the suspensions even of syllables, the juxtapositions even of parts of phrases, which he intends."[59] Howe cites this essay and Olson's poetry as influences upon her own work and her engagement with Dickinson.[60] She takes from his model of "FIELD COMPOSITION" a sense that "the page [is] an open field—words on it an instant fusion of hearing and seeing."[61] If her readings of Dickinson emphasize seeing over hearing, she is far from alone in responding to Olson in this way. Partly because of Olson's own graphical innovations in "Projective Verse" and his poems, many experimentalists writing in his wake read the poetic "field" as a primarily visual figure for the space of the page across which inscription plays.

By reworking Olson's field poetics, Howe and other experimentalists of the Language generation provide a powerful vocabulary for seeing Dickinson's manuscripts. Without this vocabulary, holograph images of Dickinson's writing would not have occasioned such a complex, energetic response in the past few decades. In this sense, our Emily Dickinson can seem as much a product of postwar and contemporary approaches to visible text as of the time when she wrote. As I discuss below, recent poetic responses to Dickinson extend this vocabulary of visible writing into the technological environments of the twenty-first century.

KNOW I KNOW, FEEL I FEEL: POETIC AFFECT AS LYRIC EXEMPTION

Dickinson's note about the affective intensity of "the look of the words" anticipates recent habits of affective response to her handwriting. Her poems also distinguish poetic affect from rationalism through the same possessive rhetoric evident in recent scholarship. When an emotion possesses her, Dickinson responds to apprehensions about viewing herself and others as objects. The more common view of emotions as things we possess (i.e., *my* feelings) also appears in many of the poems, but this configuration lacks the poetic solace available when feelings instead possess us. In an 1885 letter to Helen Hunt Jackson, the poet writes: "Take all away from me, but leave me Ecstasy / And I am richer then, than all my Fellow men. / Is it becoming me to dwell so wealthily / When at my very Door are those possessing more, / In abject poverty?" (F 1671C). By attempting to read ecstasy as an object possessed, Dickinson hazards contradiction. In the closing lines, the equation of material and spiritual wealth yields uncertainty as to whether "those / possessing more, / In abject poverty" are materially wealthy but spiritually impoverished or vice versa. Either way, Dickinson's own forms of "wealth," both spiritual and material, become the objects of a guilty conscience isolated behind the "Door" of a dual privilege. By contrast, "I cannot dance upon my Toes" is an earlier poem that explores the artistic productivity engendered when affect possesses us. It begins by casting people as objects, materials for a dance. Dickinson explores the dancer's objecthood by setting up a distinction between a certain kind of knowledge and a possessive model of affect:

I cannot dance opon my Toes –
No Man instructed me –

But oftentimes, among my mind,
A Glee possesseth me,

That had I Ballet Knowledge –
Would put itself abroad
In Pirouette to blanch a Troupe –
Or lay a Prima, mad,

And though I had no Gown of Gauze –
No Ringlet, to my Hair,
Nor hopped for Audiences – like Birds –
One Claw opon the air –

Nor tossed my shape in Eider Balls,
Nor rolled on wheels of snow
Till I was out of sight, in sound,
The House encore me so –

Nor any know I know the Art
I mention – easy – Here –
Nor any Placard boast me –
It's full as Opera – (*F* 381B)

These lines dance around questions of what the speaker does or does not
possess and what does or does not possess her. In the opening stanza, she
attributes her inability to dance not to a dispositional shortcoming but
to the fact that no "Man" has instructed her, a tellingly gendered phrasing.
This stanza ends with the poem's only affirmative instance of possession,
where the speaker becomes an object possessed: "A Glee possesseth me."
The following three stanzas unfold in the subjunctive, describing how
the speaker lacks proper equipment to express the glee that possesses her.
Dickinson sent Thomas Wentworth Higginson this poem in 1862, and
Karen Jackson Ford sees the claim to lack instruction by a "Man" as a defi-
ant response to Higginson's suggestion that she write more conventionally.
However, when Ford claims that "Dickinson's speaker takes on the gauze
and ringlets of the ballerina and then dances with a vengeance to prevent
herself from actually becoming the ballerina," she overlooks the subjunctive bracket running from the start of the second stanza through the start

of the fifth. Through this counterfactual description of what she would do "had [she] Ballet Knowledge," Dickinson does not "control and dispel" the demands of poetic convention, as Ford argues, but explores the relation between artistic know-how and the impulse to express the feelings that possess us.[62] In fact, the poem states that if she had the knowledge she lacks, she would *not* need to assume "gauze and ringlets" to deliver an affecting performance.[63] The poem reports feeling possessed by glee and describes all that the speaker does not herself possess that would enable her to express this glee artistically—first and foremost, a certain kind of knowledge.

Within the poem, affect and knowledge emerge through different ratios of possession. While the former "possesseth me," the latter is something I might possess or not. As she becomes an object possessed by affect, the speaker also loses her expressive agency: if she possessed "Ballet Knowledge," it would be the glee, not her, that "put[s] itself abroad" in dance. The rest of the second stanza clarifies an incompatibility between affect and the knowledge that would enable its expression. If the speaker knew ballet, the self-expression of her glee would still "blanch a Troupe" and "lay a Prima, mad," whether with envy or scandal. Madness stands apart from clear thinking, so it might deprive the Prima of the ballet knowledge she possesses. More adequate knowledge enables the expression of feelings, but the resulting art passes the buck of affective possession, of a feeling that possesses you and renders you passive and incapable of voicing it through careful craft. Expressive dance (and the poetic dance for which it stands) thus functions as an interface between the affective intensities that take hold of us and the technical knowledge that would seem to enable artistic expression but that is destabilized by strong feeling.

Instead of pursuing this circularity, the uninstructed speaker continues in the subjunctive, listing the props and affectations she would not need in order to captivate an audience, had she ballet knowledge. From the "Gown of Gauze" to the "Eider Balls," these absent props are superfluous, yet they can make "The House encore [her] so" powerfully that she is "out of sight, in sound." The speaker may be "out of sight" in the literal sense, behind a curtain, since the house cheers for an encore, but the synesthesia and enthusiasm also suggest reading "out of sight" as "exceedingly good." While the possession of objects such as a gown may help to express a glee that had otherwise remained inwardly felt, it is the possession or lack of knowledge that remains paramount. "Ballet Knowledge" can mean "knowledge of ballet" but also "balletic knowledge," knowledge imbued with the grace of dance. The final stanza supports this reading as it returns to knowledge and its relation to "Art." If the speaker had "Ballet Knowledge" but possessed none of the props aforementioned,

> Nor any know I know the Art
> I mention – easy – Here
> Nor any Placard boast me –
> It's full as Opera –

The second line here recalls the earlier tension between possessing artistic knowledge and being possessed by a feeling whose intensity threatens to disrupt the know-how needed to express it. Does the poem merely "mention – easy" the art of putting glee abroad, or does it perform that glee as a poetic dance? If the latter, then the poem does not simply "mention" an art but exemplifies it. However, the second and fourth lines of this final stanza do not rhyme as expected, suggesting a lack of grace that enables the poem to undercut its own status as poetic, as art. Perhaps this is the inevitable result of the tension Dickinson sets up between technical skills (like rhyming or pirouetting) and affective intensity. Meanwhile, the penultimate negative conditional, "Nor any know I know the Art," gets linked with the poem's primary conditional, "had I Ballet Knowledge," and with the main predicate of all the poem's conditions: "It's full as Opera." Where the first stanza situates knowledge in a context of instruction and learning—and of Higginson's patriarchal authority—the final stanza addresses what it might mean for the dancer's audience to know (or not) that she knows (or does not know) the Art by which her glee would put itself abroad.

This involvement of the audience motivates a few exchanges through which knowledge and affect remain in opposition. We can see what it means for an audience to know a performer knows her Art; most audiences of a virtuoso would. And an uneducated spectator at the same event might *not* know the performer knows her Art. We also can see how an audience might know a performer does *not* know her Art, as in a poor performance. Speaking in the subjunctive, however, the poem offers a false alternative: "Nor any know I know the Art" is counterfactual not just because no one knows our speaker knows the Art she mentions, but also because she does *not* in fact know it. She says so in the first stanza. What, then, would it mean for an audience not to know a performer does not know her Art?

This phantom formula of doubled ignorance, "Nor any know I [do not] know," provides an analogue for the expressive aspirations the poem voices. It aspires not to show technical know-how but to express a feeling to an audience. If the speaker's glee is incompatible with the knowledge that would enable her to express it, then perhaps its successful expression would register for the audience not as knowing but as feeling. The false negation of "know I know" insinuates an affective formulation, "feel I feel," as the ideal expressive destination of the glee that possesses the speaker. The poem opposes feeling to knowledge through the rhetoric of possession, and it thereby imagines a scene of expressive performance in which making an audience "feel I feel" glee would be as simple as folding my own lack of knowledge upon their own so that "they do not know I do not know" the Art by which my glee would express itself. This rhetoric of reciprocity also supports theorists' efforts to distinguish affect from knowledge and rational thought. Simon Jarvis puts it thus: "What grounds my being is not reflection, knowing that I know, thinking that I think, and it is thus not at all anything emptied of affectivity; it is rather the primordial fact of affectivity itself in so far as I am affect . . . the feeling which *I − am.*"[64] Whether in Dickinson's poetry or today's theories of affect, the doubled "feel I feel" displaces rationalist models of subjective reflection figured by "know I know."

At the close of "I cannot dance upon my Toes," Dickinson abandons the subjunctives and counterfactuals through which she imagines her ballet. The poem ends with the most direct statement to appear since stanza one: "It's full as Opera −." Up to this point, the poem develops a concatenation of negatively coded "o" sounds—no, nor, knowledge, opon—from which it at last flies free in the gleeful affirmation of the final word, "Opera." But what is full as opera? The penultimate stanza mentions an applauding "House," so perhaps "It's" a full opera house. The metonym again insinuates a *who* in place of an *it,* the quintessential motion of Dickinson's possessive model of affect. Yet the imagined full house seems improbable for a poem about the difficulty of expressing emotion. At the opening, Dickinson locates glee "among my mind," not among others. Both the first stanza and the final line appear outside the bracket of subjunctive fantasy, drawing a link from "among my mind" to the closing assertion. The strange preposition usage insinuates a latent sociality in even those feelings presumed inexpressible, such that what we feel in our minds might in fact be shared among them. Glossing "It's full as Opera," then, we could understand "it" not as the "House" but as "glee" itself, which after all is the referent of the poem's only other form of "it," the "itself" in the fifth line. By this view, "It's full as Opera" indicates not a full house but a feeling of joy as full-throated as opera, despite the difficulties of expressing it. Recalling us to the latent sociality of "among" rather than the solitary inwardness of "within," the closing line takes solace in the prospect of shared feeling.[65] Through its discourse of possession and lack, the poem addresses a problem of inadequate knowledge and, instead of seeking instruction, suggests that powerful feeling can supplement the rationalized techniques of craft through which we normally expect to share our feelings with others.

Already in Dickinson's poems themselves, affective life is figured through the labile rhetoric of possession. A feeling might be something we possess, or it might possess us. This changeable rhetoric of affective possession binds our feelings together with the objects that arouse and embody them. This rhetoric also configures the oppositions and exchanges between affective experience and rational knowledge. In the poem just discussed, powerful feeling is distinct from and threatens to disrupt the technical knowledge apparently needed for artistic expression, but the poem tries to imagine how affect might prove capable of communicating itself without becoming subsumed into such knowledge, subordinated to the stagecraft of artistic technique. Such exchanges between affect and knowledge unfold across Dickinson's writings and in the scholarly discourse about them. The transition from one to the other can be seen in the archival management of "Safe in their Alabaster Chambers" (F 124), one of her most widely appreciated poems.

Dickinson began writing "Safe in their Alabaster Chambers" in the summer of 1859. Over the next few years, different versions of it circulated among her family, friends, and the general public. It appeared anonymously in the *Springfield Republican* on March 1, 1862, likely sent by Susan Dickinson. The poet included a later version in her first letter to Higginson, dated April 15, 1862, these lines about the dead accompanying her famous query "if my verse is alive." In the previous year, she and Susan had exchanged multiple letters discussing revisions of the second stanza. The surviving variations and correspondence have become a lively corner of Dickinson's archive. Analysis of the correspondence about this and other poems indicates that Emily sent Susan poems not merely as favors but to get advice.[66] However, such analyses often open more questions than they answer, and the rhetoric of uncertainty that builds up around the archival materials has come to seem an end in itself. For instance, many have mentioned that, in one note Susan sent Emily about the poem, she wrote "Pony Express" on the verso of the stationery, where an address would normally go. Most who describe this detail say little about its meaning. Martha Nell Smith views it as "characterizing [Susan's] prompt response to the poem" and claims that it "also appears to refer to her sometimes acting as courier between Dickinson and the printers."[67] But she provides no evidence that Susan's note, undated as it is, came quickly after a note from Emily, nor does the note say anything about printers. Smith's readings are plausible but far from definitive. Likewise, Franklin's variorum edition refers to the "Pony Express" but does not hazard an explanation.[68] Why mention it at all? Virginia Jackson would claim that pointing out such details enables Dickinson's readers to fetishize her manuscripts and detach them from broader historical contexts.

Instead of seeking to resituate these within the reading practices and generic norms of Dickinson's time, I want to ask how the poems themselves might anticipate this impulse to mention without interpreting. I hold no brief for those who mention the "Pony Express" without explaining it, nor can I offer any better interpretation. Rather, I mean to explore the significance of this reiterated gesture of indicating a detail visible in the archive of holograph images even (or especially) when the detail does not contribute knowledge about the text or its author. In fact, "Safe in their Alabaster Chambers" and the accompanying correspondence concern themselves with just such questions of visibility and (not) knowing.

Largely unchanged during revisions, the first strophe deploys images of the deceased in mausoleums to explore a conundrum about sentience, time, and loss:

Safe in their Alabaster Chambers –
Untouched by Morning –
And untouched by noon –
Sleep the meek members of the Resurrection,
Rafter of Satin and Roof of Stone – (F 124F)[69]

The elect await resurrection, shielded from the changing slants of light that mark "morning" and "noon." Yet "untouched" also bears affective connotations; just as they are visually shielded from markers of time, the elect are emotionally unmoved by worldly events. This gloss invites a play on "morning" as an alternate of "mourning." The departed are "untouched by mourning," unconcerned with the grief of their survivors. Interment seals them from visual indices of time and from affective contact with the living, a conflation of visual and affective access that the possessive readings of Dickinson continue to replicate. The despair of the bereft is here (as always, like everything) a matter of indifference to the dead, for they cannot know it. The play on "untouched" also figures affective response as tactile, death as a loss of physical contact, a kind of contact that readers seek to restore by framing Dickinson's pen strokes as a material record of her gestures.

The early version of the poem's second strophe situates the oblivious-ness and seclusion of death alongside images of a livelier but equally irra-tional natural world. The contrast reveals latent tensions in the poem's ideas about affect and the knowingness of the natural world:

Light laughs the breeze
In her Castle above them –
Babbles the Bee in a stolid Ear,
Pipe the sweet Birds in ignorant cadence –
Ah, what sagacity perished here! (F 124B)

The natural world displays lighthearted indifference to the dead, under-cutting their supposed wisdom and importance. The kingdom of heaven may not be theirs but the breeze's, since she has a castle. This laughing breeze, the babbling bee, and the piping birds indicate that, in the face of human mortality, the natural world goes on gaily as usual. If the bees and birds produce an "ignorant cadence," senseless of their own mortality, then this surely deserves no more derision than the obliviousness of the dead, their "stolid" indifference to natural beauty. The ignorance of nature seems not so much a fault as a suggestion that the joke is on the dead. What sagacity perished here? Perhaps none at all, if the elect were comparably stolid in life. The cloistered ignorance of the dead meets its match in the aestheticized ignorance of nature.

This conceit of dueling ignorance proves untenable, though. An indif-ferent natural world provides no grounding for the trenchant closing line; the irony threatens to fall flat. If natural ignorance grants the birds and the breeze a happy indifference to death, how can they also laugh knowingly, even tauntingly, at the mortal scene below? Laughter marks a relation of interested, or at least amused, attention. Dickinson's figures of natural

ignorance encounter a common limitation of prosopopoeia: how can
nature evince total indifference to human affairs while also turning to face
us with a seemingly interested gaze? In the name of a better closing, Dickinson must exchange these figures of natural ignorance for something that
will ground her dark humor. Her first revision appears in a note to Susan,
probably sent in 1861:

> Grand go the Years – in the Crescent – above them –
> Worlds scoop their Arcs –
> And Firmaments – row –
> Diadems – drop – and Doges – surrender –
> Soundless as dots – on a Disc of snow – (F 124C)

These lines move between cosmic indifference (grand years, arcing worlds)
and *memento mori* on a human scale (dropped diadems, surrendering
Doges). The strophe closes in the breathless space where the first version failed to linger, with the "soundless" indifference of nature, this time
figured synesthetically as silent dots. In this later version, as Anne-Lise
François puts it, "the telescoping of celestial revolutions recasts the naïve
pastoral otium of the 1859 variant as the playground of centuries, where
the expanse of power is indistinguishable from its squandering."[70] In place
of birdsong's celebratory indifference, nature assumes a frigid aestheticism
more akin to Wallace Stevens's idea of "the nothing that is."[71] Both Stevens's
snowman and the soundless dots draw energy from seeming paradox, one
ontological and the other synesthetic, without abandoning a cold indifference as Dickinson's earlier version of the second stanza had done. Dickinson thus exchanges a verse about natural indifference for one that performs
it. Dickinson's galloping dactyls help to clarify her grammar, but the addition of dashes and removal of other punctuation splinter the syntax. The
effect is more intense in the original note to Susan from which we have the
revision, where the first line and others are split:

> Grand go the Years – in the
> Crescent – above them –
> Worlds scoop their Arcs –
> And Firmaments – row –
> Diadems – drop – and Doges –
> Surrender –

Soundless as dots – on a
Disc of Snow –[72]

Perhaps a function of the narrow stationery and perhaps a deliberate
choice, this form makes the syntactic and visual texture of the poem still
more fragmentary. Franklin's variorum does not render these line breaks,
though his notes mention them. If his lineation were proper, one might
expect to find it in the version copied into Fascicle 10, but there the lines
break as above. The later copy sent to Higginson contains even more line
breaks, and Franklin does render these in his edition. His regularizations
make it easy to overlook the fragmentation and syntactic ambiguity rendered by the increasing numbers of dashes and line breaks. These enable
a poetic language less intent upon providing knowledge. The dashes and
line breaks generate a paratactic rhythm that is content not to make strong
assertions but that helps the reader to respond affectively, whereas the earlier version had advanced incisive irony. Here and throughout the archive,
we cannot always say whether a line break is intended or is merely the
result of Dickinson's unusually large script in relation to often narrow stationery. This uncertainty about the most basic formal questions might
count as an important factor in the experience of reading Dickinson, and
many studies that privilege the manuscripts make it so.

The unsure status of these formal elements suits their function in this
poem quite well. In the revised second strophe, dashes and line breaks
interrupt and block the problematic argument about natural ignorance
provided in the earlier version. Through them, the poem exchanges an
explicit reflection on nature's happy ignorance for a more direct performance of ignorance, foisted onto the reader who cannot quite navigate
these formal, syntactic, and conceptual instabilities. The new stanza may
seem to assert knowledge in the form of a cosmic history, but its oblique
descriptions exceed the scope of interpretation and are thus likelier to
occasion nihilism than functional knowledge.

The increasingly fragmentary versions of the second stanza are most
powerfully rendered in "Emily Dickinson Writing a Poem," a section of
the *Dickinson Electronic Archives* edited by Smith and Lara Vetter in which
holograph images of the poem and attendant correspondence are presented together with transcriptions and editorial commentary. The transcriptions preserve the lineation of the originals, offering an alternative to
Franklin's regularized treatment. In both this digital collection and Smith's

earlier reading of "Safe in their Alabaster Chambers" in *Rowing in Eden,* the archive underscores the importance of Susan Dickinson's contributions to the poetic production now attributed exclusively to Emily. The online collection includes a note from Susan that asks, "Has girl read Republican," which possibly, but not certainly, refers to the printing of "Safe in their Alabaster Chambers" there. In the note, Susan complains, "It takes as long to start our Fleet as the Burnside," a reference to General Burnside's siege of Roanoke Island some weeks before the poem appeared in the *Republican.*[73] Smith and Vetter believe the plural possessive refers to the poems and, by attributing joint ownership, confirms Susan's importance as a contributor to a poetic process often attributed to a solitary genius.

The digital holograph images in "Emily Dickinson Writing a Poem" (re)construct a poetic collaboration between Emily and Susan, a collaboration that unfolds through an affectively charged rhetoric of pleasure and withdrawal. Whether issued on paper or online, efforts to reedit subsets of the Dickinson archive often aim to give readers an impression of more direct contact with the materials in their original, generally handwritten form.[74] Within such projects, holograph images function as naked evidence of a historically actual event in language. Those who follow Franklin, Michaels, and others in believing a text must be transferrable from one physical form to another, however, might see editions such as "Emily Dickinson Writing a Poem" as polemical manipulations of source documents, aimed not at a neutral presentation of historical evidence but at the advancement of a certain point. Whether it is found in the documents or conjured through their manipulation, "Emily Dickinson Writing a Poem" portrays Emily Dickinson's revisions of the second stanza as directly responsive to Susan's pleasure, displeasure, and indifference. The collection's first holograph is the note from Susan about "our Fleet," which strangely begins with indifference: "Never mind Emily – to-morrow will do just as well – Don't bother." The editors follow this with a note in which Emily offers a revision of the second stanza (*F* 124C) and hopes that "perhaps this verse would please you better – Sue." The women's discussion of the poem is based upon reciprocal affective investment, a desire to know the other's feelings and to give pleasure, rather than a less personal aesthetic or intellectual standard of judgment. Susan disparages the new version in affective terms, rather than critically parsing its flaws: "I am not suited dear Emily with the second verse – . . . it does not go with the ghostly shimmer of the first verse as well as the other one – It just occurs to me that the first

verse is complete in itself it needs no other, and can't be coupled – . . .
I always go to the fire and get warm after thinking of it, but I never *can*
again."[75] Higginson would concur in 1890, calling the revision a "too daring condensation" but admitting "it strikes a note too fine to be lost."[76] For
Susan, the problem is not that Dickinson's "condensation" leaves her cold.
Quite the opposite. The first stanza hauntingly depicts the eternal cold of
death, making her seek the warmth of a fire even as it reminds her that she
"never *can*" find full solace in such earthly comforts.[77] The digital collection traces Emily's continued efforts to make a suitable "peer" for the first
stanza. She would produce two more versions, both included in Fascicle
10 and plainly inferior to the first two attempts. "Emily Dickinson Writing
a Poem" provides evidence of Emily sending Susan one of these two later
versions. The letter begins not with a salutation but with an abrupt question: "Is *this frostier?*"

> Springs – shake the Sills –
> But – the Echoes – stiffen –
> Hoar – is the Window – and
> numb – the Door –
> Tribes of Eclipse – in Tents
> of Marble –
> Staples of Ages – have
> buckled – there –

I again follow the transcription provided in "Emily Dickinson Writing a
Poem," rather than the regularized Franklin version, to show that the frag-
menting effects of line breaks and dashes intensify. Dickinson hews to the
imagery of the tomb that telegraphs mortal cold so powerfully in the first
strophe; instead of contrasting this imagery with natural vitality, phrases
like "Tribes of Eclipse – in Tents / of Marble" align the two previously
opposed movements of the poem. The holograph image of this strophe
places it in a note to Susan, meaning that it provides more evidence of
the women's exchanges about poetry, but as poetry these lines prove still
more obscure, an even more "daring condensation" than the first revision.
As a closer reading of the holographic context reveals, Emily does not even
expect Susan to understand the new strophe in the usual sense.

   Following this third version of the strophe, Emily writes, "Dear Sue –
Your praise is good – to me – because I *know* it *knows* – and *suppose* – it

means –." What praise? Susan has said she is "not suited" with earlier revisions. Explaining why she counts Susan's response as praise, Emily produces the same structure of doubled knowledge we saw in the poem written later that year, "I cannot dance opon my Toes." That poem's epistemic *mise-en-abyme*, "Nor any know I know the Art," clarifies Emily's view that Susan's (non)praise is good "because I *know* it *knows* – and *suppose* – it *means*." In both cases, a statement about reciprocal knowledge proves deflationary in relation to knowledge itself. Reading Susan's response, Emily may "*know* it *knows*" better than to praise imperfect verse. Knowledge enables discernment but not pleasure. Given the goals of satisfaction that guide their exchange (Susan's with the revisions and Emily's with Susan's responses), the "frosty" praise is good not as critique but as something else. I argued above that the affective doubling of "feel I feel" haunts the ballet poem's dual negation of knowledge in "Nor any know I know," proffering affect as a basis of expression. In the same way, "*suppose* – it *means*" lingers inscrutably behind "*know* it *knows*," an alternative to the distancing gestures of critical discernment. If Emily understood Susan's cold response as knowing, she would know better than to call it praise but would view it at face value. If the response is "good" as "praise," this is because Emily can "*suppose* – it *means*," can draw its meaning (as praise) not from its own knowing criticism but from vaguer supposition. In response to the critical judgment that disrupts the exchange of pleasures between her and Susan, Emily brackets the exchange of knowledge in an overly tidy formulation, "I *know* it *knows*." (One thinks of the chain-smoker's demurral, "I *know* it's killing me, but still . . . .") She prefers a less knowing attitude about their correspondence, so she can "*suppose*" Susan's feedback "*means*" to offer the affective support of praise. As in "I cannot dance opon my Toes," Dickinson situates affect in competition with the isolating asceticism of knowledge. She articulates alternative structures of reciprocity—"*know* it *knows*" running aground on "*suppose* – it *means*," just as "know I know" insinuates "feel I feel." She thus posits affective experience as an alternative to the exchanges of knowledge endemic to poetry criticism and ballet training.

These efforts to produce a suitable second strophe for "Safe in their Alabaster Chambers" stage exchanges between knowledge and affective response. Through parataxis, synesthesia, and oblique phrasing, the revisions dramatize nature's irrationality instead of thematizing it, as the first version had done. Dickinson thus pursues a "*frostier*" affective texture even as she chooses to "*suppose*" Susan's criticism "*means*" to offer praise, rather than the stark knowledge of a shortcoming. Those reading holograph images of this poem and attendant correspondence pursue analogous strategies of lyric exemption that seek affective rather than rational truths. In their preface to "Emily Dickinson Writing a Poem," Smith and Vetter provide a bulleted list of questions, and these indicate the uncertainty that seeing holograph images arouses:

- What is the identity of "Safe in their Alabaster Chambers," the poem under Emily and Susan's consideration here?
- Is it a two-stanza poem with four different second stanzas, as their writings and the contemporaneous printing that Dickinson saw suggest?
- Is it a three-stanza poem, as rendered in its 1890 posthumous printing? Or is it in fact five one-stanza poems?

Viewing Dickinson's holographs does not provide clearer textual information but heightens uncertainties about what even constitutes "the poem under . . . consideration." For Smith and Vetter, the point is not to look closely at the holograph images and decide which stanzas count as part of the poem; the point is that, as one looks closely at the handwritten sources, the very object of one's critical attention seems to shift and dissolve into an unmanageably complex textual web. Through this process, a text that had seemed open to rational reading strategies instead becomes a matrix for discourses of affective experience, where affect is continually distinguished from critical thinking. Like many of Dickinson's poems, the readings that follow this trajectory describe affect through a language of possession whose applicability to both persons and things, to Dickinson and her material possessions, sustains their rhetoric of lyric exemption.

Indeed, deploying these images in this way enables Smith and Vetter to inscribe an affective exchange between Susan and Emily alongside their poetic collaboration. If the images provide a basis for some kind of knowledge, it is knowledge about an affective bond. Smith and Vetter refer to Emily's correspondent as "Susan Huntington Gilbert Dickinson, her most beloved friend and sister-in-law." The superlative wording insinuates the common—and reasonable—belief that the women were lovers, not simply "beloved" family. They later reinforce this suggestion, describing Emily's attentiveness to "her beloved's advice," but they also assert that their collection primarily chronicles "a literary dialogue." The difference between affective rhetoric and critical thinking has powerful effects here. It enables the editors, like many others, to insinuate a love affair without making a decisive claim one way or another, but it also suggests that affective life itself remains traceless or inscrutable so often that we should not expect a decisive record of this relationship to emerge. Like Smith's earlier writings on this poem, "Emily Dickinson Writing a Poem" recuperates the poet's quotidian exchanges with Susan in order to challenge "literary traditions that have drawn sharp distinctions between 'poetic' and 'domestic' subjects." Just as the holograph images upset such distinctions, the editors coordinate a language of affect to disrupt fixations upon whether Emily and Susan were lovers: the lesson and the point of affect's difference from rational knowledge is that such emotional bonds elude evidentiary verification. Dickinson's poems anticipate this distinction between affect and rationality that today's apparatuses for viewing holograph images seem to sustain.

THE LOOK OF THE WORDS—

Emily Dickinson's earliest editors already saw the importance of visual contact with her handwriting, often for the same reasons that the manuscripts interest readers today. The first holograph image appears in *Poems: Second Series* (1891), edited by Higginson and Mabel Loomis Todd. The volume's "frontispiece" is a four-page "fac-simile" of the poem now known as F 325, "There came a day – at Summer's full –." Typeset versions of this poem had appeared the previous year in *Scribner's* and in the first series of *Poems*. In his variorum edition of 1998, Franklin claims that Todd and Higginson included the facsimile to verify the spelling of a word in the seventh line, "soul," which Susan Dickinson repeatedly and erroneously changed to "sail" (see Franklin's commentary following F 325D). Maybe so, but Todd's preface makes no mention of the disputed word. The second series of *Poems* appeared after the unexpected popularity of the first, and the facsimile may therefore have served more to evoke a cult of personality around the poet than to provide textual information. In her preface, Todd describes how Dickinson's handwriting gradually changed from "the delicate, running Italian hand of our elder gentlewomen" to a "bolder and more abrupt style" in which "each letter stood distinct and separate from its fellows." She is the first to assert that "the handwriting makes it possible to arrange the poems with general chronological accuracy," a method Thomas Johnson and Theodora Van Wagenen Ward carry forward in Johnson's variorum edition of 1955.[78] Though Johnson considers this dating "conjectural,"

Ward lists the graphological minutia by which she determines the year of each manuscript. At times, her methods are so meticulous as to appear implausible. For instance, she claims the signal characteristic of manuscripts from 1868 is that the *T* no longer appears "triangular" but is "made with two strokes." Elsewhere Ward deploys affective language, as when the handwriting of 1862 appears "Less agitated than in 1861."[79] As Todd predicted, these methods made it possible to establish a rough chronology, but Franklin achieves more accuracy because he supplements handwriting analysis with closer attention to the material condition of the manuscripts: changes in stationery, traces of binding and other organizational methods, and even stains on the pages.

Johnson's access to many of the manuscripts was limited. He examined those now held at Amherst College on only two occasions, though he was given photostats of them. As a technology for viewing Dickinson's manuscripts, the photostat machine complicated and, indeed, limited the kinds of knowledge available to Johnson. Yes, without photostats, his work might have proved impossible. But without them, he might also have sought freer access to the originals and thereby arrived at a more accurate chronology. Already in Johnson's edition of 1955, then, electronic equipment for viewing Dickinson's handwriting does not simply provide more textual information but actually complicates and defers the kinds of knowledge an editor might hope to find in the images such equipment enables us to produce.

The earliest holograph images do not primarily serve evidentiary purposes but establish a mood. As Todd writes in her 1891 preface, "the effect" of seeing Dickinson's handwriting "is exceedingly quaint and strong."[80] The 1891 facsimile's placement as a frontispiece may stem partly from convenience, since such divisions enabled printers to separate complex images from the normal typesetting of the main text. But as a frontispiece, the facsimile also appears paratextual, decorative. It presents in manuscript a poem already published in the first volume of *Poems* (1890), so it directs readers' attention not into the volume it adorns but, like other frontispieces, outward to the broader contexts of literary production. In her 1894 introduction to the first edition of the poet's letters, Todd again calls the handwriting "quaint" and laments that "the coldness of print destroys that elusive charm" evident in the handwriting.[81] Hot script, rather than cold print, telegraphs the affective charge of the language. Two of the three holograph images in *Letters* (1894) are excerpts. The first of these presents the opening paragraph of a letter and the signature, eliding the intervening paragraphs visible in the adjacent typeset version.[82] Such images distort the texts to which they correspond instead of facilitating more detailed textual knowledge. Other early facsimiles likewise do not present the handwriting for informational purposes but to provide a sense of affective proximity with the living poet. The linkage between affective reading and holograph images that so powerfully informs recent approaches to Dickinson therefore seems nascent in even the earliest editions.

Seeing Dickinson's handwriting does not provide textual information so much as it refers us, like a photographic portrait, to the poet's personhood and her mortality. Whatever equipment produces a holograph image, it indexes the affective life of a woman who has died but not passed away. As Roland Barthes tells us, the photograph's essence is the indexical claim, "that-has-been," that leaves a visible trace in the present of a person who, at the time of the photograph, was already going to die: "If the photograph then becomes horrible, it is because it certifies . . . that the corpse is alive, as *corpse:* it is the living image of a dead thing."[83] We might say the same of Emily Dickinson's holographic corpus, burned not into ashes as the poet asked but into the photosensitive materials through which her handwriting proliferates. For viewers of this corpus, "the look of the words" indexes the actual having-been of Dickinson's expressive act, the event of inscription. These textual remains enable affectively rich experiences of contact with the poet. For instance, Ward's short note on the handwriting of Dickinson's final year, 1886, marks no difference from the previous year, but it anticipates the poet's death: "Large, loose, and badly formed, showing physical weakness."[84] Like the face in any photograph, the late manuscripts mark, for Ward, the place of a person who was going to die. Where a holograph image stands in for the gesturing body of Dickinson, this substitution switches between the objective and subjective, between the *it* and the *who* of affective experience. For Barthes, to be photographed "represents that very subtle moment when . . . I am neither subject nor object but a subject who feels he is becoming an object: I then experience a micro-version of death . . . I am truly becoming a specter."[85] This objective–subjective switching produces a visual rendition of facticity per se, of the "that-has-been," rather than of any particular fact.

By far the most visually distinctive element of Dickinson's oeuvre is the dash. More than anything else, the dash is what makes Dickinson poems *look* like Dickinson poems. Few are the poets whose work can be recognized at a glance, without reading a word, and fewer still those whose entire output is so closely linked with a single mark of punctuation.[86] Any visually oriented response to Dickinson must, of course, grapple with the dash, both with its graphical dimension and with its semantic, syntactic, and aural effects. This mark's shifty grammar—its powers of apposition, elision, fragmentation, and trailing conclusion—shapes her distinctive syntax. The dash also supports affective readings, since it signifies the wordlessness of strong feelings, their tendency to disrupt language. The impressive variety of critical claims about the dash and about strategies for printing it condenses many of the concerns that structure broader discussions of affect in her work and of the visibility of her handwriting.

When casual readers understand themselves to see a dash Emily Dickinson wrote, what they most likely see is a dash placed according to Johnson's 1955 variorum or Franklin's 1998 variorum. Both editors made numerous decisions about which marks in the manuscripts would be presented as dashes and which would not. Even where the presence of a dash is undisputed, the editions differ. Johnson's presented itself as undoing the regularizing work of earlier editions and thus offering versions of the poems closer to their state in manuscript; this edition's long em dashes emphasize its resistance to the pressures of convention, its fidelity to the poet's unusual punctuation. By contrast, Franklin's en dashes are more restrained, but they also approximate more closely the relative length of most dashes in the manuscripts. I probably am not the only one who, at first, found Franklin's en dashes infuriating because they seemed to diminish such a graphically important component of Dickinson's work. Of course, I had come to know that work through the em dashes of the Johnson edition, which now seem awkwardly elongated. The Dickinson poems scattered online and in lesser-known collections generally draw upon Franklin or Johnson, carrying forward their decisions as to dashes. Where an en dash or em dash cannot be rendered, readers might find one or two hyphens in its place, another technological turn in the visual life of Dickinson's corpus. These considerations may seem fairly arcane and diacritical, but they fundamentally inform what a reader sees when she sees a Dickinson poem. Together, they form a poetics, quite literally setting the conditions of appearance for the poems. The increasingly common experience of seeing a Dickinson holograph image reshapes these questions about how a Dickinson poem—and especially her dash—should look to a reader.

Critics and editors rendering the dash in type tend to view it as informational, as fundamentally transferrable from one instance of the text to

another. But as images of Dickinson's holographs have become more prev-
alent, even renditions of the dash in type have freighted it with the kinds
of preverbal, affective expressivity those most invested in the handwritten
dash tend to privilege. In his review of Johnson's variorum, Spicer reads
the dashes as a kind of score for vocal inflection. Under Johnson's hand,
Spicer believes, the typeset dash provides a kind of prosodic information
that had previously remained unavailable in print. Smith and Paul Crumbley have likewise published typeset dashes in which they find some kind
of metatextual information. In *Rowing in Eden,* Smith sets the poet's dashes
at varying elevations and lengths and describes the slanted dashes as a
"rhetorical notation" that imbues a poem with comedy or irony. But when
Smith does not say why a given dash is longer or shorter, higher or lower
on the line, these variations evince a gratuitous attention to the visual complexity of the holograph. Indeed, she often places "dashes" in scare quotes
to indicate that the marks rendered as dashes in print encompass a wide
variety of complex lines. She reads some such marks as performing the
"visual mimicry" of expressionist drawing, as when the downward slant
of a line accompanies the arrival of a fly.[87] Crumbley, in the most extensive
study of Dickinson's dash, provides typeset dashes that vary in length,
slope, and elevation. He reads the dashes and related marks as "highly
nuanced visual signals intimately linked to Dickinson's experiments with
poetic voice," arguing that they support her "strategy for investing readers
with the authority to challenge social determination of linguistic content."[88] Crumbley is far from alone in believing Dickinson's dashes can be
organized into a typology in which different kinds of dashes signify differently. The problem is that everyone making this informational claim has
developed a different typology and a different strategy for representing the
dashes without adequately explaining why their scheme is better than the
others. Stranger still, as we saw above, Howe and Werner include in their
criticism what the latter calls "diplomatic transcriptions" that set the letters
of a poem in type but include hand-drawn dashes to match the length,
elevation, slope, and complex shape of each handwritten dash. Despite the
peculiarity of this hybrid approach, Howe and Werner make surprisingly
few concrete claims about the information these hand-drawn imitations
might provide. If the poet's handwritten lettering can be fairly transferred
into type, why not her dashes too? The handwritten condition of the dash
is a point of fixation because it posits a degree-zero of inscription's gestur-
ological dimension, its tracing the actual embodied movement of the poet.

Because of its form, the dash refers inscription to its basis in gesture, the motion of writing a line. Even the period, a simpler and commoner punctuation, seems derivative because it is made through the nonmovement of an implement that marks by moving; the linearity of so many of Dickinson's handwritten periods attests as much. To draw a short line is the most basic move the writing hand can make. We have already seen how readers from Todd onward involve Dickinson's handwriting in various informational systems. For Todd, the handwriting transmits information about class and culture. For many, her changing script indicates the date of a given manuscript. Still others read her eccentric late script as a sign of failing eyesight and unsteady hands. In each case, the handwriting provides metatextual information that no typeset version could convey. The most basic information of this sort is a gestural tracing epitomized by the single stroke of the dash: her hand moved thus. By tracing embodied gestures, the dash also refers us to the noninformational register of affect. Many who wish to distinguish affect from emotion hold that the former describes the intercalation of feelings with the bodily sensations arousing them and the gestures expressing them. As the trace of an expressive motion, the dash is affective because it indexes the visceral, preverbal register at which affective experience unfolds. Here in this singular dash of her pencil, the poet's felt passion emerged from her body onto the page. Reading strategies that reserve a place for the handwritten dash—that conceive the holograph not as textual but as graphical, as the tracing of embodied movements— thereby situate affect as exempt from full articulation in the rational system of print. Whether the smooth motions of such gestures can ever be fairly captured by electronics that discretize and quantify them remains a major contention among phenomenologists and media theorists.

It is tempting to claim that today's affective responses to the visible Dickinson resulted largely from the past few decades' technological innovations. No doubt the personal computer and the internet, as well as the digital scanner and camera, laser printer, and imaging software, have made it dramatically easier to see holograph images and other visible traces of Dickinson. But to set up the silicon chip and its dependent technologies as strongly determining our present interpretive habits is to ignore a continuum of electrical equipment that reaches back to the first editions of Dickinson and indeed to the technoscientific imaginary of her poems themselves. In 1890, when Lavinia Dickinson paid for the plates that would print the first series of her sister's *Poems,* she paid for a process called electrotype.

This process uses a weak electrical current to deposit metal onto an object. After taking a mold from the typeset *forme,* printers would have brushed the mold with graphite or another conductive coating, wired it to a cathode, and immersed it in an electrolyte solution together with a block of copper connected to an anode. Copper ions would slowly coat the mold, and the resulting copper plate could be separated from the mold and used for printing. From the very first editions, then, electrical equipment set the conditions for Dickinson's publications and for the making-visible of her writing. Even if the influence of such equipment upon our reading practices has intensified in recent decades, any narrative of such technological determination should begin early.

The idea of electricity as a medium of affect also is not new. Scholars have amply described the reading of electronic technologies emergent in Dickinson's era as analogues for the affective life that hides away in the nervous system, itself increasingly understood in that period as an electrochemical system.[89] This trope on electricity as a link between the human body and its affective intensities appears often in Dickinson's poems. She describes "Hope" as an "electric adjunct" to the muscle called the heart, and a "Curl" of "Electric Hair" recalls a distant lover (F 1424, F 628). In later poems, electricity materializes emotion, as when "Distain" is an "electric gale" of the soul or when "fervor" manifests as an "electric Oar" to propel us (F 1448, F 1631). When critics write of the "electric passion" that coursed through the poet's life, though they may partly be responding to the recent electrification of her archive, they deploy a figure already familiar to Dickinson herself.[90] Dickinson's own electrical figures for affective response make it possible to put some pressure on expectations about the encounter with holograph images today. One can readily see why scholars handling Dickinson's manuscripts might find the experience affectively intense, given the importance and fragility of such documents. More surprising is the relative ease with which such affective intensity has been transposed onto the encounter with printed facsimiles and, more recently, with holograph images glowing on a screen. But already in the language of Dickinson's time, the notion of electricity as embodied emotion invites a reduction of any affectively intense experience to its electrochemical basis in the nervous system. The "look of the words" can set off a storm in the sensory nerves that strikes to the affective interior Dickinson called "the soul." The fact that such a storm can begin with a computer screen should perhaps seem less surprising. In multiple senses, the visual encounter with Dickinson's handwriting has always been electric.

ABSENCE: ITS POETICS AND POLITICS

Just as a photograph can mark the having-been of a person now departed, Dickinson's holograph images hold the place of a certain absence. Her pen strokes would arouse less interest if she had overseen conventional publication of the poems. The record of such a process would clarify her attitude about typesetting, which many consider a violent deformation of her work. Would Dickinson have titled her poems before publication? How would she have formatted those written on oddly shaped paper? Would she have made final selections among variant words, or refused to do so? Of course, such questions simplify the poet's complex attitudes about publicity and printing. But imagining the forms their answers would take, had she overseen normal publication, suggests something important about today's holograph images. To whatever extent these record the affectively intense gestures of poetic expression, they also mark the *absence* of a more mundane textual practice, that of transcribing and editing and proofreading for the press—a practice Dickinson never undertook. Yes, she had her own processes of transcribing and editing, of binding and distribution, but her manuscripts matter most where they leave undecided questions that her instructions to a publisher might easily have settled. These humdrum activities of typesetting contrast with the affectively charged scribblings so many see in the holograph images, but the nonoccurrence of these more placid activities enables the images to stand in their place and thereby informs the epistemic conditions of the Dickinson corpus. Today's holograph images, then, are not solely determined by recent technological development, nor do they simply record an expressive event of inscription. They also enable readers to rehearse the lack of a textual practice that might have stabilized the epistemic terrain of Dickinson's writing. The manuscript images do not clarify knowledge but function *an*epistemically, providing sites of affective response where critical inquiry falls short. Affective interpretation is thus positioned as a collaborative other of knowledge, not subject to critique but still able to underwrite claims of certainty, and its lability in relation to knowledge further constructs what will and will not count as true of Dickinson's work, what will and will not be seen.

Such constructions inform not only the epistemic but also the political meanings of the texts they work over. In particular, the visible mutilations of Dickinson's letters and manuscripts organize the poet's availability to identity-political appropriations. Her damaged papers occasion a melancholic discourse of loss and recovery that recruits her for feminist and queer readings. The writings most frequently mutilated are those that express love for her sister in-law, Susan. Smith claims these erasures "track the path of an early editor suppressing evidence of Emily Dickinson's intense engagement with her primary audience, the woman to whom she sent more writings than she did to any other."[91] Like Franklin and others, Smith attributes most of the mutilations to Todd, who, in addition to editing the early collections, was the mistress of Susan's husband, Austin Dickinson. Perhaps Todd resented her lover's wife enough to expunge her from the archive, or perhaps she intended to protect Emily and Susan from suspicions that they were a bit *too* intimate. That a single textual mutilation can be read as either aggressive or protective should indicate the epistemic unsteadiness of the damaged texts. Someone, for some reason, was invested enough in these texts to attempt effacing some of them, but the image of a mutilated text can reflect neither who did the damage nor why. Instead, the mutilations and the half-legible texts they overwrite indicate that these dual traces of affective life—the poem itself and its violent overwriting— must remain visible yet unverifiable, legible yet obscure, open to affective reading but not to claims of fact.

The most affecting and politically salient holograph images are the most difficult to read, such as the fascicle copy of the poem now known as F 5, "One Sister have I in our house–." This poem as it appears in Fascicle 2 in Franklin's *Manuscript Books* and elsewhere is entirely scribbled over in heavy ink pen, likely an attempt to hide the poem's declaration of affection for Susan (Figure 4). In 1858, Emily Dickinson sent Susan a separate copy of the poem, and this copy remains undamaged. Martha Dickinson Bianchi, Susan's daughter, vindicates her mother by offering this as the dedicatory poem of her edition, *The Single Hound* (1914), titling it "To Sue." The undamaged manuscript of the poem is pasted into Bianchi's copy of her edition. Even without this undamaged copy, Franklin can discern most of the writing beneath the scribbling in the fascicle. His variorum edition thus includes two versions of the poem with minor variances. Only by virtue of the text's persistent legibility under erasure can we perceive the affection for Susan it expresses. We could not read a complete erasure as an attack upon that affection, since we could not read through to the object of its occlusion. Only as *not* erased, as still visible, do the mutilated texts become legible as damaged traces of emotional contact between the two women. We may heed Smith's call to "imagine scars on the body of Dickinson's writings," but we produce this narrative of loss only by reading as lost those texts that remain minimally legible.[92] Although Franklin's *Manuscript Books* provides a transcription of the cancelled poem, Smith's online essay about the mutilations does not. Throughout Smith's collection, clicking the "transcription" link above a holograph image normally invokes a typed transcription alongside the image. Clicking for a transcription of the damaged holograph of "One Sister," however, yields the note "[Completely Inked Over]." One cannot readily make out the writing behind the cancellations when viewing the holograph in Franklin's *Manuscript Books* or in Smith's digital collection. Franklin seems to have managed with first-hand access to the manuscript, but Smith does not reproduce his transcription here. For Smith and others who linger over the mutilated scraps of Dickinson's corpus, the display of the damaged text *as illegible,* a strategy to interrupt rather than facilitate reading, takes precedence over efforts to see the poem more clearly. The visual encounter with such illegibility affirms the tracelessness of affective life, the impossibility of any writing that would adequately transcribe feeling itself.

This structure of visible invisibility shapes the discussion of a possible lesbian relationship between Emily and Susan. Throughout her career, Smith has argued that the intensity of the two women's relationship has been minimized in order to cast Emily Dickinson as an isolated genius or a jilted heterosexual spinster. Smith does not, however, definitively claim Susan and Emily had sex, nor even that erotic desire primarily structured their exchanges: "More interesting to me than the suppression of the erotic is the suppression of evidence of mental, compatriot intellect between two women."[93] Attempts to inscribe the relationship between Emily and Susan within a lesbian framework might be seen as strategies to occlude the greater scandal of two women thinking together. If Smith admits that the "erasures are at least in part discourses of desire," we still cannot say whether the effaced language originally described a sexual exchange or, rather, this desire is our own. In important ways, the debate about Susan and Emily has not clarified their relationship so much as it has produced more uncertainty, encouraging a discourse of affective response understood as distinct from evidentiary inquiry.

Figure 4. The poem as it appears in Fascicle 2 has been defaced with ink pen. Courtesy of Amherst College Archives and Special Collections.

Uncertainties about the nature of Emily's relationship with Susan are thus as much a result of affect's inchoate material traces as of the epistemology of the closet. We are often reminded that women of Dickinson's era enjoyed sentimental bonds with other women without always considering these romantic. Fair enough, but this fact does not clarify Emily's particular bond with Susan, which may have been more or less typical for the time. Such blockages of affective knowledge appear also where nonnormative sexuality is not directly at stake: we are equally helpless to name the Dark Lady or the Fair Youth of Shakespeare's sonnets. Damaged or not, the material record cannot provide sure evidence of a writer's affective life, since feelings do not organize themselves according to evidentiary logic. Just try proving that you love someone. We can never surely say what affective currents may have coursed through a life as it was lived, nor can we expect writing to tell us where queer desire might emerge, where a thought or act of love might have made its unrecorded claim. Still, there is a long tradition of imputing strong emotion to women, especially artistic women, so opposing affect to rationality enables a misogynistic coding by which women's affective experiences mean they are less rational, more passively at the mercy of their feelings. Howe, Smith, and others intend that affective readings of Dickinson should provide a more social, worldly alternative to feminist readings of her as a paragon of cloistered genius, but we remain uncertain how to read Dickinson's gender outside this double bind: isolation or affective engagement. Such uncertainty informs discussions of Dickinson's sexuality and thereby her political significance. That we should *not* know whether Emily and Susan had sex makes them at least as available to queer reading as a more definitive record of sexual acts would. Through the discourse of the closet, the rhetoric of uncertainty (and particularly the rhetoric of affect as an alternative to exposure) has become central to contemporary feminist and queer reading strategies. We might say of Dickinson's current political meanings what Eve Sedgwick says of her own queer epistemology: "A point . . . is *not to know* how far its insights and projects are generalizable."[94] The configuration of affect as a reading strategy exempt from rationalist constraints draws upon and energizes a queer politics that seeks alternatives to the protocols of unveiling that demand verifiable evidence of a writer's identity and acts. No wonder the readers most interested in Dickinson's affective life accord a special significance to the damaged and semilegible manuscripts. They sustain a visual economy of loss and recovery that provides an indefinitely extensive way to unravel the political meanings of her work.

POETIC POSSESSIONS OF
DICKINSON TODAY

Janet Holmes's *The Ms of My Kin* likewise explores the logic of loss and erasure to recruit Dickinson for political purposes. The book contains strategic erasures of poems Dickinson wrote during 1861–1862, the outset of the U.S. Civil War and the years of Dickinson's transition to a more experimental mode. The title itself is an erasure: *The ~~Poems~~ of ~~Emily~~ ~~Dickinson~~.* The words Holmes leaves behind constitute poems even sparer and more elliptical than Dickinson's—poems that now appear to address recent U.S. wars in the Middle East. In a short preface, Holmes writes, "I owe the project to the invitation of its epigraph." This epigraph already signals the interplay of absence and inscription the book sets going. It is the opening two lines of the poem now known as *F* 184: "If it had no pencil, / Would it try mine —" Reflecting on possession, on a pencil one has or lacks, these words introduce the switching between subjects and objects that the language of possession so often occasions, for in the same breath that Dickinson mentions her pencil, she writes of a person as an object, an "it" rather than a "he," "she," or "they." Among other effects, the impersonal pronoun elides the person's gender. The full text of the poem does nothing to clarify the gender of "it," although "it" does become involved in a discourse of sexual possession and objecthood:

> If it had no pencil,
> Would it try mine –
> Worn – now – and *dull* – sweet,
> Writing much to thee.
> If it had no word –
> Would it make the Daisy,
> Most as big as I was –
> When it plucked me?

Franklin's variorum edition notes that the verse was "signed 'Emily' and sent to Samuel Bowles," the poet's friend and editor of the *Springfield Republican.* Knowing this does not dispel the strange effect of the impersonal pronoun. Virginia Jackson, arguing that this is not a lyric poem, views the pronoun as "grammatically impersonal but rhetorically intimate."[95] Perhaps the strange usage alone bespeaks a certain intimacy, but the impulse to see unconventional rhetoric as a sign of intimacy with a correspondent, not as a basic dispensation of writing poetry, tells us more about Jackson's perspective than about Dickinson's text. Indeed, the introduction of a personal pronoun in line four suggests some rhetorical distinction between the "it" that might borrow a pencil and the "thee" to whom Dickinson has written "much" and is writing now. The sexually suggestive reminiscence in the final line, "it plucked me," again sets some distance between "it" and Samuel Bowles, whom most consider the poet's platonic friend. Framing this text as primarily a letter to Bowles thus elides much of its interest as a poem. Today its opening lines are as much an epigraph of Holmes's book as they are an appeal to Bowles to write—perhaps more. Since none of us is Samuel Bowles, viewing these lines primarily as a letter to him constructs a dyadic social bond whose interior remains more or less illegible to any reader today. (Did they have sex after all? Were they in love? No one can say.) Holmes detaches the first two lines from the rest of the poem. She reads them not as inviting Bowles to write but as inviting her to erase. Compared with a historicist perspective intent on clarifying the past, Holmes's appropriative misreading is not only the more poetically generative but also the more reflective of the interpretive habits Dickinson's writings have helped to develop.

The technological backgrounds of Holmes's epigraph also occasion a discourse of absence and loss. Franklin's variorum tells us that the surviving manuscript of the poem was "pinned around the stub of a pencil" (218). As Jackson notes, neither Johnson's edition nor Franklin's could "include the pencil."[96] But why does Jackson point out its exclusion? This gesture contributes to the same economy of absence and recovery, of visibility and occlusion, that her rhetoric of archival specification would hope to displace with the fullness of historical details. What would even count as "including the pencil"? Jackson's book contains a photograph of the manuscript page with the stub of pencil sitting atop it and the two original straight pins reinserted through the paper. However, the photograph is in high-contrast black and white, which is good for handwriting but so dramatically shaded that I did not at first recognize the pencil at all. Its shadow is darker than the pencil itself; together, they look more like an ink stain or mark of redaction than an object sitting on the paper. Further, it is unclear whether Jackson has correctly reinserted the pins through the paper. The Amherst College digital collection includes high-quality color images of the page, the pencil, and a small archival card through which the pins have been stuck. On this card, the pins are labeled "Upper pin" and "Lower pin." Jackson accordingly reinserts them at the top- and bottom-left corners of the unfolded sheet. Creases in the page and a third pair of holes on opposing side margins halfway down the page indicate the page was folded into thirds horizontally before the pencil was used as a vertical axis for rolling the paper. In this case, the "Upper" pin Jackson has reinserted in the top-left corner of the page would originally have pierced the lower corners of the paper (which was folded superior to the upper corners prior to the rolling and pinning), and vice versa. I make these suggestions without having examined the manuscript in person, for I am less interested in its material actuality than in the interplay of information and uncertainty that unfolds across its photographic appearances.

Together, these images register multiple kinds of absence that elude historical efforts to repair and reassemble them. The most comprehensive online collection, the *Emily Dickinson Archive,* includes a high-definition color photograph of the manuscript page but does not show or mention the pencil or the two pins. These objects, already absent from the variorum editions, are absent in a new sense here. A computerized comparison indicates the *Emily Dickinson Archive* image is identical to that in the Amherst digital archive; the page was not rephotographed. Yet the Amherst image of the pencil and pins is omitted from *Emily Dickinson Archive.* The electronic image of the manuscript makes visible the multiple pinholes at each of the six sites where the page is punctured. These and the small tears between certain holes resulted from Dickinson's rolling the page around the pencil multiple times before inserting the pins, or else from various attempts to reassemble the enclosure over the years, or both. No matter how careful the effort to place the pins in their original positions, the artifact remains marked and damaged by an absence that informs how we know it: the manuscript and its images are quite literally full of holes. Last but not least, the Amherst image of the pencil shows a detail not evident in Jackson's photograph but particularly salient for Holmes's work: the nub of pencil Dickinson sent Bowles has no eraser attached. (The first pencil with attached eraser had been patented only three years before Dickinson sent this note.) If Dickinson's redirected address

invites Holmes to "try" her pencil, lacking one of her own, then Holmes's
project of erasure still relies upon a technology not already present in the
archive. Holmes instead "tries" the markings of Dickinson's pencil against
the revelatory gestures of their partial effacement by twenty-first century
technologies. Throughout *The Ms of My Kin*, the voices and circumstances of U.S.
military activity after 9/11 seem to emerge from the erased texts of Dickinson's poems. Most illustrative of all is Holmes's erasure of *F* 247, "The
Lamp burns sure – within –." Here is the full text of that poem as it appears
in Franklin's variorum:

> The Lamp burns sure – within –
> Tho' Serfs – supply the Oil –
> It matters not the busy Wick –
> At her phosphoric toil!
>
> The Slave – forgets – to fill –
> The Lamp – burns golden – on –
> Unconscious that the oil is out –
> As that the Slave – is gone.

This poem is rarely discussed in the criticism. Jonathan Morse views it
as "withdrawing from the outward servility of body to the inward queen-
dom of mind."[97] Through this withdrawal, Domhnall Mitchell argues, "the
focus of the poem is on the lamp, but not on the servant who supplies
it with oil."[98] Mitchell links this elision of labor with the truism that a
poet must make her verses seem effortlessly constructed. These allegorical
readings are warranted, but the poem's fundamental occasion is the negli-
gence and departure of a forced laborer. The one who does not supply
the lamp with oil is the one who thereby *does* supply the poem with its
central tension. Remarkably, the word "slave" does not appear in any other
Dickinson poem, nor does it appear in any of her letters. Mitchell claims,
"the 'Serf' introduces a historical distance that makes the 'Slave' Roman
not southern, universal not particular."[99] It strains credulity, however, to
think that a prolific American poet's sole mention of a "slave," in a poem
written months after the start of the Civil War, connotes anything other
than an African American slave who has escaped. After Holmes's erasure,
however, the poem indicates a different kind of absence:

It matters

that the oil
is gone.[100]

Holmes began the book in 2003, as the United States invaded Iraq. Her endnote lists "occasional speakers of the poems" including "those piloting aircraft on 9/11; U.S. President George W. Bush; Osama bin Laden;" and others.[101] Of course, Dickinson cannot have foreseen recent wars and left descriptions of them hidden in her poems. Holmes's erasure conjures the voices she mentions. Erasure would seem an obtuse way to assert that Dickinson's work contains the elements of a discourse about war and politics. Why not just read the originals for that discourse? Faith Barrett argues that "the further disintegration of a centralized lyric speaker in Holmes's erasures signals the obliteration of the lyric self in the face of 150 years of violent conflicts fought by American soldiers."[102] But Holmes does not simply reject the lyric voice; she redacts Dickinson's and constructs others. She challenges the notion of lyrical voicing by performing its impossibility. Her erasures reveal voices that cannot be there, ghosts of the future lurking in Dickinson's language.

The words Holmes erases inform the historical thinking of her work as powerfully as those she leaves behind. Most striking are the erasures of "Slave" in the second stanza. If the lamp stays lit even though "the Slave – is gone," perhaps the sentiment is abolitionist. We do not need Holmes for such a reading, but in order to construct a poem about present-day wars, Holmes must erase the words most decisively linking Dickinson's poem with the war of its time. It matters that the "Slave" is gone, in other words, because Holmes must redact the language of slavery in order to render a discourse of war for oil. Poetic erasure's relation to historical memory becomes clearest when erasure appears prior to inscription, opening the possibility of inscription through reference to its outside. John Paul Ricco understands erasure this way in an essay about drawing: "Drawing presupposes erasure as originary absence.... Drawing is then to be understood as an inscriptive amnesia, the forgetting ... of erasure. Ironically, then, erasing is the remembering of this absence within drawing, it is what helps drawing remember its own forgetting."[103] Holmes's erasure remembers in multiple senses. It produces a false memory in which Dickinson's poems had always cryptically described modern wars in the Middle East; it reminds us of the Civil War, to which Dickinson more directly alludes, but only by redacting words associated with this earlier conflict; and finally, per Ricco, it reminds us of the forgetting that supplements inscription itself.

Dickinson's poem and Holmes's erasure offer complementary responses to the economies of scarcity and absence that govern historical memory. The metaphor that closes Dickinson's first stanza already performs a wishful forgetting: "Tho' Serfs supply the oil / it matters not the busy Wick / at her phosphoric toil." The wick forgets not only the conditions of labor supplying the oil but also the logic of scarcity by which a real wick would burn out if not resupplied. Holmes traces over this paramnesia by herself forgetting Dickinson's fantasy of plenitude. From "it matters not," Holmes draws "it matters," an erasure that reinscribes a scarcity of matter, a mattering, over Dickinson's wishful thinking. Dickinson's side of this copula imagines a wick for which it "matters not" that the slave might flee and let the oil run out. On the other side appear the prevailing material, legal, aesthetic, and indeed political logics of scarcity that have governed Dickinson's reception. These include the almost accidental series of events that caused Dickinson's poems to be published and celebrated rather than burned and forgotten, the institutional strategies of possession and preservation that still determine who can and cannot see which Dickinson manuscripts, the queer epistemologies that have written and unwritten Dickinson's possible erotic engagement with women, and indeed the rhetoric of "telling it slant" by which many readers cast Dickinson's verse as epistemically obscure. Despite Dickinson's faith in plenty, those who read her do so in a world where "It matters that the oil is gone," where this and other forms of scarcity govern who lives and dies, what we see and do not. We read, moreover, in a world where Dickinson's works remain subject to damage, loss, erasure. Holmes's poem produces its historical parallax by erasing the language of forced labor, of slavery, that otherwise marks Dickinson's text as historical—language that the original poem itself seeks to displace into a metaphysics where such forced work and the violence it occasions might not be necessary.

Occasionally I have referred to Holmes's technique as redaction instead of erasure. "Redaction" emphasizes the disciplinary function of her process and, perhaps, of any exchange between visibility and effacement.

Barrett begins to make this point when she links Holmes's book to "the censoring language from intelligence reports," an affiliation also evident in Travis Macdonald's *The O Mission Repo,* a strategic erasure of *The 9/11 Commission Report.* Holmes's redactions bring her source texts into relation with the state at formal and aesthetic, not only thematic, registers. As a redaction, Holmes's book subjects Dickinson's poems to the same suppression that polices the difference between visibility and secrecy in the conduct of war today. In the end, though, Holmes lets us off easy. Reading between Dickinson's poem and Holmes's redaction positions them as negative images of each other: the first articulates a fantasy of plenitude to elide anxieties about forced labor, and the second erases in order to reassert the logic of scarcity that governs modern exigencies of labor, of energy, and of reading. However, reading between them also affirms the endurance of interpretation and meaning, of legibility over and against the negative force of erasure. Perhaps we cannot perceive eraseability as a condition of writing until there is "nothing to see here," as the police say. What if the combined image of Dickinson's work and Holmes's were a blank page? What if Holmes had gone into the archive at Amherst or Harvard and actually erased a Dickinson manuscript? We might come to see Holmes herself among the terrorists who voice her book, but if so, she would have rendered the same terror of the blank that she already helps us to think through.

Recent poetic responses to Dickinson not only illuminate the visual politics of inscription and erasure, as Holmes does, but also explore the politics of affect. Michael Magee's *My Angie Dickinson* exemplifies a young genre called flarf, whose exponents understand poetry's political significance in primarily affective terms. Magee's book, like most flarf, was produced through "Google sculpting," in which the poet searches the internet for certain phrases—in Magee's case, "Angie Dickinson" and selected words from Dickinson's poetry—and then culls language from the results to "sculpt" a poem. Parroting online slang and pop idioms, flarf performs vapidness and stupidity. Instead of responding critically to popular ignorance and vulgarity, it mocks and mimes the irrational habits of mass culture. Magee emphasizes the political significance of this strategy. Expressing rueful nostalgia for the "devious" machinations of Nixon and his cohort, he avers that, by comparison, "George W. Bush is an utter dumbfucking fool achieving the same effect," continuing, "I feel compelled in the face of this to interrogate dumbness, ridiculousness, stupidity; to work undercover in the middle of it."[104] Holmes, of course, shares Magee's opinion of Bush, but she asserts critical objections instead of joining his perverse nostalgia for more intelligent villains: "This World // baffles — // Men / of // Faith slip — and / see / Evidence — // in / lies —// their // Reward for this — // Empire."[105] By contrast, Magee sees subversive potential in miming the stupidities of public culture.

Flarfists often frame the political significance of their methods in affective terms. Rick Snyder describes Nada Gordon's flarf as recording "the pathos of being an anonymous individual in the emotional chaos of a post-9/11 world."[106] As in the possessions of Dickinson, pathos gives flarf writers a way of responding to a loss of contact between individuals or between an individual and the social body. For Gordon, this kind of poetry might respond to political hopelessness with affectively laden rhetorical questions: "'Aren't we all a bunch of fools, and isn't that funny? and bittersweet? and fucked up?'"[107] A lyrical pathos emerges from these circuits of loss and recovery; in the discourse of flarf, this is the loss and recovery of political possibility. Magee explores how a rhetoric of poetic exemption, taken as a license to be unserious, might recuperate the "low" public discourse online, suggesting that, by "sculpting" such language to look like Dickinson poems, he can either elevate it to the condition of poetry or challenge the idea of poetic value itself. His poetry seems *of* its political situation rather than opposed to it. The political interest of these books by Magee and Holmes may lie in exploring the apparent impossibility of setting up a decisive critical distance from the present situation, of untangling the mutual reliances that make a poet feel complicit in everything from Google's commoditization of language, to the history of slavery in the United States, to the same country's more recent colonial projects in the Middle East.

Such impossible hopes may refer to something more than political critique; they might also imagine a world in which we could be possessed by our feelings and possess them too. Magee's *My Angie Dickinson* figures the impossibility of this dream:

The Mind is the most Exquisite Torture—
World rejoices—It's a very Good Thing—
Infibulation—closing penisvagina—with suture—
Or ring—

The Heart—is a Handout—for Dicussion
Island's faltering tourist—Trade—
Remembrance of The Stomach—is shortened.—
Real name: Angeline.[108]

Even without reference to Dickinson's handwriting, the visual economy of her work remains important here. Magee sprinkles his poems with dashes and impromptu capitalization: if you stand back and squint, they look very much like Dickinson poems. His preference for em dashes harkens to Johnson's midcentury efforts to make the poems look less like conventional verse and more like they look in the manuscripts. The first stanza of this poem turns from platitude to shock. Infibulation might cause or stem from mental torture, or it could be affiliated with the world rejoicing, perhaps as a grim distraction from the tortures of mental life. The genital "closing" might remind us that any closure between a world rejoicing and a mind that tortures itself would prove this macabre. The second stanza makes analogous transitions. Just as the mind is a seat of torture rather than of knowledge, the heart is not a self-certain interiority but a "handout." It is given away freely, like charity, or it is ephemeral as a classroom exercise. Either way, the heart is no longer that organ most surely mine, most closely linked with my feelings. We can pass it around and discuss it as we discuss an island's tourism or a "remembrance of the stomach," perhaps of an especially good or bad meal. The *who* of affective experience becomes an *it*, ready for distribution. As Jennifer Ashton puts it, "what's striking about Flarf . . . is just how impersonal the personal presence in the poem is."[109] Through the intimate public space of the internet, matters of the heart turn to matters of consumption, commodity, and economy— more objective modes of possession.

The final line again explores the objective–subjective ambivalence that results from a possessive rhetoric of affect. "Real name: Angeline." By articulating her full name, Magee suggests that even our attachment to a Hollywood star provides shorthand ("Angie") for a spiritual elevation ("Angeline") that may be latent within his poems in the same way Emily Dickinson is. In the foreword to his book, Magee explains his strange conflation of the two Dickinsons:

Why *Angie* Dickinson? Most obviously to disrupt some of the pieties around Emily Dickinson's work that I don't believe have served her poems very well. (As an example, I would note the rarely mentioned fact that Emily Dickinson is one of the funniest poets ever.) Then too, Angie Dickinson is a sort of Zelig figure in American popular culture (and in particular on the internet).[110]

Nonetheless, one remains apprehensive about the contingent linkage of the two names. One suspects that, if the actress had been named Edna Pound, we might now have a book called *The Edna Pound Era*—no doubt a longer book. Born Angeline Brown, the actress adopted her first husband's surname, so her link to Emily is a sheer coincidence, albeit shaped by a patriarchal kinship system. To call her a "Zelig figure" is to elide the equally patriarchal history by which a talented actress was cast in increasingly sexualized roles that exploited her body more than her skill as a performer. Angie Dickinson's star sign in postclassical Hollywood was subject to the same deflationary trajectory that flarf itself explores.[111] The title of Magee's book and the poems themselves mention Angie, never Emily. Glancing at the cover, which bears drawings of Angie in various roles, one would have to know Howe's work to catch the reference to this other, more famous Dickinson. The interplay of the two names constructs a system of insider knowledge based upon false expectations of obliviousness. Among the few who actually read such books as Magee's, the insider knowledge is all but assured; almost no one in Magee's audience will not know Howe's work, let alone Emily Dickinson's.

I wonder not so much why Angie as why *Emily* Dickinson. Some see the postpolitical collage techniques of flarf as pursuing "the conscious erasure of self or ego."[112] In that case, what motivates Magee's possessive attachment to an individual writer and to Emily Dickinson in particular? If Angie serves "to disrupt some of the pieties around" Emily, then what pieties might Emily herself disrupt or invoke? For one thing, she provides literary prestige to a genre normally more at home with cult celebrities and canned phrases. Insomuch as flarf tarries with a parodic distance from its cultural referents, the book's apophatic claim upon Emily Dickinson might seem to mock or displace the earnest rhetoric of possession that shapes *My Emily Dickinson* and other affective responses. But by emphasizing the visual texture of Dickinson's poems, Magee reads alongside rather than against those who, like Howe, address her works as visual productions. As Magee's appropriation of the possessive title suggests, he extends the familiar enchainment of possession, affection, and lyric exemptions from rationalism.

By the time of Magee's writing and of Holmes's, an approach to Dickinson arrived at almost by chance, through Howe's appropriation of the poet's own possessive phrasing, has combined with a series of technological developments to encourage a form of affective interpretation that has broad consequences for the afterlives of lyric expression today and that now seems all but inevitable as a perspective on Dickinson in particular. The next chapter explores the rhetoric of lyric exemption in terms of this conflicted view of history, as both a continuous causative process and the product of certain seemingly random events. There, an exchange between Gertrude Stein and Jackson Mac Low opens a discourse on chance. Chance connotes cold impersonality rather than passionate intimacy, but it has proved attractive to American poets and their readers partly because, like affect, it offers counterintuitive ways to engage with electronic information systems.

2 Chance

Gertrude Stein, Jackson Mac Low, and A Million Random Digits

Near the end of his life, the experimental poet Jackson Mac Low used equipment involved in the design of nuclear weapons to rewrite several works by Gertrude Stein. The resulting *Stein* series comprises over 160 poems, about two dozen of which are published. To compose each poem, Mac Low selected and rearranged words from a given Stein text through a combination of chance operations, deterministic procedures, and his own decisions. To perform many of the chance operations, Mac Low did not employ such familiar equipment as dice or the flip of a coin. Instead, he used a thick reference volume titled *A Million Random Digits with 100,000 Normal Deviates*, which he had discovered in the 1950s and used occasionally since then. This strange book consists almost exclusively of a vast table of random numerals. It was produced in the late 1940s by the RAND Corporation, a military think tank, at the request of scientists at the Los Alamos Scientific Laboratory. The scientists needed a large set of random numbers for a new mathematics called the Monte Carlo method, a method developed to aid the design of thermonuclear weapons. In short, the so-called RAND book was created at the dawn of the Cold War to facilitate the production of the deadliest weapons ever built. The foreword to the online edition emphasizes that the book has since found more peaceful uses among "statisticians, physicists, polltakers, market analysts," and others, but its notoriety stems from its role in nuclear armament.[1] Given its military origin, *A Million Random Digits* is an extraordinarily unlikely piece of equipment for the experimental writer's toolkit. Indeed, Mac Low was an avowed pacifist. Prior to the *Stein* series, he most often performed chance operations with a deck of cards or the *I Ching*; the former recalls his longstanding interest in the aesthetics of play and the latter his Buddhist mysticism.[2] By using the RAND book to create the *Stein* poems, he embraces chance as a mode of lyric exemption that tests the limits of historicist interpretation.

Approached through its chance-operational equipment, the *Stein* series also clarifies the stakes of Stein's reception among postwar and contemporary poets. Many experimental writers continue to view Stein as a vital precursor. However, the *Stein* series does not simply posit a predictable avant-garde lineage, from modernist experimentalism to Language poetry to conceptual writing. Instead, the series emerges at the intersection of multiple literary, political, and technological histories, including the history of nuclear war, all of them shaped by an idea of randomness as a mode of *historical regard.* In other words, Mac Low's use of the RAND book underscores how randomness organizes the interface between a literary text and its historical contexts, including the literary traditions to which it responds and contributes. Randomness thus provides a supplement to strictly rationalist approaches to history, a role it plays in a surprising variety of intellectual fields. To think about randomness challenges normal ideas about causation and agency for both writer and reader, so it complicates historical sense making and attributions of literary value. When things happen just by chance, sequential narratives appear ungrounded, destabilized by a sense of arbitrariness.[3] As we shall see, Mac Low and Stein share an interest in this relation between randomness and historical meaning, but they understand it quite differently. Stein does not agonize the implications of randomness for historical thought but takes a pragmatic approach to the question. Through the RAND book, by contrast, Mac Low takes randomness as an aesthetic strategy to disrupt historical meanings and value judgments, to see these as absurdly accidental yet still bearing consequences worth our attention. Despite the disruptive effects of randomized procedures, his *Stein* poems linger with ideas of personal expression and literary value, ideas not typically affiliated with chance-operational texts. Mac Low thus complicates the normal opposition between procedural writing and more traditional lyric modes. His approach in fact exemplifies a broader lineage of experimental writers, reaching back to Stein, who have explored similar questions about the contingency of historical meaning. His poems provide insight into the rewards of interpreting randomly generated texts, showing us how the play of chance disrupts historicist reading strategies.

As an avenue for interpreting the *Stein* series, *A Million Random Digits* offers productive cues for new media poetics. The previous chapter argued that Emily Dickinson's respondents use a family of "information technologies" to differentiate poetry from the information such devices ostensibly make visible, strategies that might be seen as poetic *détournements* of essentially rational machines. By contrast, others view electronics as involving irrational processes from the start. Mac Low's use of the RAND book demonstrates that digital composition can be playful and unpredictable, not just informatical. Likewise, as I argue below, scientists working in America's wartime cybernetics labs did not see themselves as merely designing better logic engines. Rather, they used the RAND book and similar equipment to address slippery questions about uncertainty, noise, paradox, and randomness, thereby situating disorder and contingency as core problems for what has become modern computer science. Today, we more commonly understand computers as mechanisms for gathering, manipulating, and presenting positively given information, but this view elides their inventors' interest in the others of rationalism and the frequency with which these devices inspire less knowledge-oriented responses such as frustration, uncertainty, disorientation, confusion, and play. *A Million Random Digits* underscores the largely unrecognized ways these less rational, less instrumental modes of thought shape our interactions with digital media. A fully developed digital poetics might account for and even celebrate such noninformatical interactions with computers while questioning the economies of knowledge such criticism itself produces. As a model of randomness in digital poetry, the RAND book points Mac Low and his readers in that direction. Meanwhile, the rhetoric of "digital" media often implies that numbers enjoy a mathematical ideality severed from any material substrate or historical context. The sheer disorder of the RAND book further suggests that its digits lack any meaningful relation to their own material history. The book's enduring power to signify historically, however, indicates that future work in digital poetics would benefit from clarifying whether and when such mathematical ideality actually holds, a question Mac Low and the RAND book help us to address.

Poets and critics often view chance operations' indifference to an author's intentions or biography as a way of bracketing historical contexts and unsettling aesthetic and political influences.[4] In this sense, chance functions as a mode of lyric exemption by disrupting the poem's situation within rational systems of causation. However, Mac Low's chance-operational

equipment also does the opposite; the RAND book insinuates an un-
comfortably violent history within the scene of writing. It thus suggests
that even a poetic practice intent on decontextualizing itself historically
emerges from specific material histories. That said, the operations the
RAND book facilitates really are random, and this randomness does as
much to shape the *Stein* poems as do the more determinate contexts from
which they emerge, such as Mac Low's fairly predictable decision to rewrite
Stein and not, say, Robert Frost. The *Stein* series brings into focus the
stakes of such a decision, making it easier to tell what Stein does and does
not share with the postwar and contemporary experimentalists who claim
her as an influence. Contrary to recent criticism that emphasizes her political and stylistic conservatism, her work pursues a more thorough departure from the coordinates of traditional lyric expressionism than even Mac
Low provides. Prior to the *Stein* series, Mac Low had used the RAND
book on other occasions and had rewritten other canonical modernists, as
in *Words nd Ends from Ez* (1989), his rewriting of Ezra Pound's *Cantos*. In
the *Stein* series, however, his use of the RAND book intersects productively with Stein's own aesthetic commitments. More broadly, his chance
operations challenge historicist interpretation: How do you render historical meaning from texts produced at random and, at the same time,
acknowledge that such randomness gives a feeling of hermeneutic weight-
lessness, as though things might easily have happened otherwise? In view
of the equipment used to produce them, the *Stein* poems illuminate the
encounter between randomness and literary interpretation, and through
them, chance emerges as a mode of lyric exemption. How this mode might
have a history, be it literary, political, or technological, is this chapter's ani-
mating question.

SOURCING STEIN

Mac Low produced each *Stein* poem according to a process he called "diastic reading," a combination of chance operations and rule-governed procedures whose product he then adjusted according to his own taste. To
describe this procedure, I will trace the production of one poem, "Stein
100: A Feather Likeness of the Justice Chair." Having chosen a Stein work
to rewrite—in this case, her famous 1914 book of prose-poetry, *Tender
Buttons: Objects, Food, Rooms*—Mac Low began by randomly selecting
from it a "seed" text, a passage that would direct the rest of the procedure.[5]
To choose a seed for "Stein 100," Mac Low drew a random number from

the RAND book and thus arrived by chance at the fifty-third paragraph of *Tender Buttons:*

> A fact is that when the direction is just like that, no more, longer, sudden and at the same time not any sofa, the main action is that without a blaming there is no custody.[6]

Mac Low then applied his predetermined procedure to this randomly chosen paragraph to produce his rewriting. Starting at the top of the larger "source" text from which the seed comes (here, *Tender Buttons*) Mac Low selected from it the first word whose first letter matches the first letter of the seed. Here the seed's first letter is *a*, and the first word in *Tender Buttons* to begin with *a* is the indefinite article, "a." Mac Low then moved to the next letter in the seed, the *f* in "fact." Since that letter begins a word, he selected the next word in *Tender Buttons* to begin, like "fact," with an *f*: "feather." Mac Low then resumed in *Tender Buttons* where he left off, now searching for a word whose second letter, like "fact," is *a*, then a word whose third letter, like "fact," is *c*, and then one whose fourth letter, like "fact," is *t*. He then moved on to the next word in the seed, "is," and sought a word in *Tender Buttons* whose first letter, like "is," is *i*. This yields "a feather table reckless gratitude it." The spelling of the randomly selected seed thus dictates which words get chosen. Here are the first three strophes of "Stein 100," which comprises eight strophes and seventy-five lines in total. I have underlined the letters that key the words to the seed above:

A feather table: reckless gratitude.
It is that-there that means best.

White the green grinding trimming thing!
The disgrace, like stripes.
More selection, slighter intention.

Rosewood stationing is use journey: curious dusty empty length.
Winged cake: the cake, the plan that neglects to make color certainly.
Time long could winter: elegant consequences monstrous.
So much and guided holders garments are—and arrangements.
Staring then that when sudden same time's necessary, that circular
    same's more necessary, not actually aching.

In the "makingway endnote" to each *Stein* poem, Mac Low calls his method "reading-through text selection," for he must literally read through Stein's text to find the words that fulfill the procedure. Unsurprisingly, "Stein 100" closely resembles its source text, *Tender Buttons*. Words like "table," "trimming," "feather," and "dusty" recall the book's domestic environ, and some of the poem's stranger constructions, such as "white" as a verb and "disgrace" as a pattern, recall Stein's own syntactic oddities. But Christopher Funkhouser and others have warned against overstating the "primacy of the database" in discussing poems that, like Mac Low's, draw words from larger source texts.[7] Indeed, when Mac Low's interventions in the procedure mark out his differences from Stein, they offer greater insight into reading procedurally generated texts. For instance, the *Stein* poems have line breaks, but *Tender Buttons* and most of their other source texts do not. The diastic procedure does not make the line breaks. Mac Low identifies them among the "personal" touches he makes after generating a poem, but they do not provide a vehicle of personal expression. Mac Low endstops every line, so the line breaks give the poems a mechanical feel that recalls the rigid procedure itself. Even the exclamation point in line three, normally a sign of expressive intensity, primarily helps readers recognize the line as a command; the phrasing is too abstruse to sound exclamatory. Here too, Mac Low differs from Stein, who famously took a sparing approach to punctuation. As Lisa Siraganian notes, the paucity of punctuation in *Tender Buttons* reflects Stein's view that "*all* diacritical marks . . . should be eliminated in order to confirm the irrelevance of the reader" in relation to textual meaning.[8] By contrast, Mac Low reintroduces punctuation and line breaks as tools to shape the reader's experience, but he deploys them in a way that makes the poems feel all the more procedurally managed. Contrary to Louis Cabri's claim that "punctuation in Mac Low's texts . . . manifests textual and authorial singularity," Mac Low's ostensibly "personal" use of punctuation and line breaks ironically suggests that a text produced through impersonal chance operations and rule-based procedures cannot necessarily become more expressive through manipulation of its formal and diacritical elements.[9] As Funkhouser writes of digital poem generation, some poets "seek to be imitative" of their source texts, but others use the procedure itself "to alter or subvert typical forms of expression."[10] Mac Low's departures from Stein accordingly help readers consider what features can make a text expressive. The *Stein* poems ask whether an expressive voice can emerge from texts written by chance operations and, if not, how we can interpret them.

Here and in later readings of the *Stein* series, I focus on the idea of personal expression because it was central to Mac Low's thoughts on his techniques and remains a key question for experimental poetry. Reflecting on his methods in the 1980s, Mac Low begins by noting that his late work often mixes chance operations with personal decisions. Soon, though, he describes his "motive for using chance" as a will to compose poems "relatively 'uncontaminated' by the composer's 'ego,'" and he avows that "it was such a relief to stop making artworks carry that burden of 'expression'!"[11] In such contexts, a poem counts as expressive if its details disclose some meaning intentionally and personally crafted by its poet. Charles Bernstein has argued that Mac Low's procedural work "refuses the normal process . . . of writing as confessional or personally expressive" in order to present "text severed from its author-ity."[12] This idea of expression remains a common object of disavowal among today's conceptual writers and related experimentalists, as in the influential anthology edited by Craig Dworkin and Kenneth Goldsmith, *Against Expression* (2011), which showcases texts by Mac Low and others who, often through chance operations, seek to escape the traditional coordinates of lyric expression. The *Stein* series, however, complicates such notional accounts of antiexpressive writing by leading readers to ask how a poem written through chance operations might nonetheless solicit expressionist modes of interpretation. For many readers, "lyric" has come to name the set of generic effects that solicit these kinds of interpretation.[13]

Not all theories of expression focus on the individual. For instance, when Theodor Adorno refers to the lyric's "principle of individuation," its personal voice and its immersion in particulars, he values this as the index not of someone's inner life but of a broader social reality. For him, "the entirety of a society . . . is manifest in the work of art," not despite but rather through its particularity.[14] Others favor a more strictly individualist theory of expression that guarantees the intentionality of the signifying act. Without this guarantee, interpretation might run aground on the impossibility of differentiating between a textual detail chosen on purpose and one that appears by chance. But social theories of expression such as Adorno's do avoid the accidentality that threatens interpretation; the ground of expression simply shifts from an agentive subject to the broader systemic orders that produce the social. In other words, for Adorno, instead of expressing one person's thoughts or feelings, the lyric expresses a social state of affairs. He notes, however, that chance disrupts this social register of expression by displacing the subject from the expressive process. In aleatoric art, "the aesthetic subject exempts itself from the burden of giving form to the contingent material it encounters, despairing of the possibility of undergirding it, and instead shifts the responsibility for its organization back to the contingent material itself."[15] Adorno rejects such art as formally incoherent, arbitrary. But we might reconsider, for if the play of chance has a structurally meaningful function in the poem's social world, then chance reflects a social reality instead of simply disrupting aesthetic forms. Such is the situation in the constellation of texts Mac Low sets up. The RAND book provides a shorthand for the technoscientific strategies of randomization that gained currency during the mid-twentieth century in fields ranging from finance to literature to physics, and it thereby confirms chance's significance within the social systems from which Mac Low's writing, and indeed Stein's, emerge. The *Stein* poems challenge us to understand not only how the institutionalizations of chance get mediated through a personal, expressive voice but also how chance sets the conditions for such expression. They exhibit the surprising persistence of individual expression that comes from its interdependence with the functions of chance commonly believed to destabilize it.

"Stein 100" may claim "slighter intention," suggesting a diminution of personal choice, but Mac Low's decisions to deviate from the rewriting procedure often improve legibility. As Peter Middleton has argued, Mac Low's procedures at first seem to frame the poems as "instances of diminished agency," but closer inspection reveals that his "poetic language is not the machine it seems."[16] With these formulations, Middleton rehearses a common but problematic opposition between free agency and mechanicity, an opposition that gets complicated by the play of chance in these poems. But Middleton is right that the traces of Mac Low's intentional decisions persist in the texts. Mac Low adds "means" in line two of "Stein 100," providing an object for "that-there" and suggesting that, if the poem "means" anything, it does so thanks to his interventions, not to the procedure. Likewise, in the sixth line, "Rosewood" paradoxically fulfills the procedure's rules while suggesting that, to be fulfilled, they must be broken. *Tender Buttons* sets the word as "rose-wood" (*TB*, 14). Mac Low encountered it while searching for a word whose eighth letter is *o*, to match "direction" in the seed text, and "rose-wood" qualified only if he counted the hyphen as a letter. But hyphens typically did not count for Mac Low. To expedite the rewriting, he often used Charles O. Hartman's computerization of his procedure, a program called DIASTEX5, and the program also does not seem to view hyphens as letters.[17] Moreover, "Stein 100" itself indicates that Mac Low treated hyphens as diacritical marks, not letters. After all, he added a hyphen in "that-there that means best," as though it were a sort of punctuation to be arranged by personal choice. Stranger still, the published version of "Stein 100" renders the word as "Rosewood," omitting the hyphen that qualified the word for selection in the first place. Perhaps Mac Low breaks the rules of his procedure at will, or perhaps he accidentally dropped the hyphen at some stage. Discussing a similar series of procedural poems, the Asymmetries, he notes that when he makes a mistake in the procedure, he views "mistake-making as an auxiliary chance operation" and retains the error. If he later scrutinizes a detail in a poem, he has "no way of knowing whether this is an accepted mistake or the result of a decision made during composition." And that is the point. The *Stein* poems likewise leave their reader unsure whether the text "means best" when Mac Low sticks to his procedure or when he breaks the rules, "either by chance or choice," and they often make it impossible to tell the difference.[18]

Such indecision forces the reader to grapple with what Jacques Derrida calls "the hermeneutic compulsion," the impulse to interpret even when one cannot distinguish probative textual details from those appearing by chance.[19] Mac Low's work underscores the interpretive difficulties that result from the play of chance in language. Derrida describes language as "one among those systems of *marks* that . . . *simultaneously* incline toward increasing the reserves of random indetermination *as well as* the capacity for coding and overcoding or, in other words, for control and self-regulation."[20] The *Stein* poems invite us to reevaluate interpretation that takes place within such systems, where the play of chance becomes indistinguishable from and interdependent with systemic control and regulation, including the regulation of hermeneutic norms in a range of fields, from atomic physics to experimental poetry.

Ironically, Mac Low sometimes downplays the role of chance in his procedures. Here is what he says about his techniques in a letter submitting some *Stein* poems to a journal:

The methods I've used since the 1960s are actually deterministic rather than chance operations in that if one uses them in the same way with the same sources and seeds they will always produce the same outputs . . . but now I always, to some extent, modify the results of the procedure, making personal decisions of many different kinds.[21]

Mac Low thus views the *Stein* poems as exploring the interface between deterministic procedures and more expressive authorial shaping. He distinguishes between procedural determinism and his "personal decisions" but repudiates chance, seeming to forget the chance operation with which his procedure begins. Yes, given the same seed and source texts, a *Stein* poem is nearly reproducible, but that "given" excludes the random selection he mentions in each "makingway endnote" as the first step of his process. Every published *Stein* poem includes such a note, and he carefully describes the RAND book whenever he uses it. When Bernstein points to "Mac Low's meticulous insistence on documenting the system, conditions, and time of each work," he indicates the poet's impulse to make his equipment and methods a part of the poetic record.[22] Mac Low's later reluctance to acknowledge the project's reliance on chance operations when describing the series as a whole suggests that, despite using a "method through which poems could mostly 'generate themselves,'" he hopes to maintain some space for personal expression.[23] As we shall see, his reluctance also indicates some trepidation about the military equipment he uses to accomplish the chance operations.

It is therefore important to attempt close readings of these randomly generated texts. Under such close attention, an expressive voice seems to emerge from the *Stein* poems. The series leaves readers unsure as to whether this hint of personal expression should be taken seriously or is simply an accidental effect of the encounter with a procedurally constructed text. For instance, a later line of "Stein 100" states, "They asked that her speech be repeated."[24] One cannot help but wonder whose speech this could mean and who "they" could be. Can "her" refer to Stein? "They" to some audience? Similarly, the opening line of "Stein 32" declares it "Pleasant to be repeating what they went to."[25] Perhaps "what they went to" refers to Stein's original work, as Mac Low and so many other postwar experimentalists "went to" Stein as a precursor. Maybe the poem finds it "pleasant to be repeating" words so often appreciated. Elsewhere in "Stein 100," the poem again seems to comment on its own project: "Imitation?—imitation is a joy gurgle."[26] Even as the poems "gurgle" with the disorder of a text arranged through an arbitrary procedure, attentive readers will encounter seemingly expressive moments of self-awareness. As Tyrus Miller puts it, Mac Low's work explores "the persistence of syntax and other forms of cohesion even in chance-generated texts."[27] Neither Mac Low's comments nor the texts themselves equip readers to decide whether the poems' apparent expressiveness is a sheer accident of the procedure, thus meriting no special attention, or a poetically significant aspect of the work. Mac Low does not help us to arbitrate this difference. Rather, this indecision itself highlights a fundamental difficulty of interpreting poems that, like Mac Low's, test the limits of traditional commitments to authorial choice and poetic expression.

The *Stein* poems ask, without necessarily answering, how one might begin to interpret procedural or chance-operational poetry without some notion of personal expression to ground the meanings that emerge from texts such as these. The lingering whisper of an expressive, self-aware voice in the *Stein* series appears as a kind of lure, inviting strategies of lyrical interpretation that the poet's procedural methods should embarrass. Such interpretation would ignore the unhappy origin of Mac Low's equipment, since the RAND book provides merely random digits. However, within a broader frame that emphasizes Mac Low's impersonal procedure and the historical contexts enabling his work, the origin of his equipment never fully recedes from view. Mac Low's published notes on the *Stein* poems offer detailed information about his equipment and writing process, but the poems themselves insinuate an expressive voice that troubles the effort to read them in terms of how they were made. Mac Low thus minimizes the importance of chance in his work not because he thinks it trivial but because he senses its power to complicate his own relation to poetic expression and to the literary and political histories through which he writes.

STEIN'S INTEREST IN CHANCE

The Stein texts Mac Low rewrites explore analogous questions about chance but to different ends. In other words, Mac Low does not simply recruit Stein for his own purposes; his rewritings clarify how chance informs her experimental style and her aesthetic theory. Mac Low most often selects words from *Tender Buttons* (1912; 1914), *A Long Gay Book* (1912; 1923), and *Pink Melon Joy* (1915; 1922). While many Stein scholars focus on the more accessible texts written before 1911 and after 1915, experimental poets who consider her an important precursor prefer what Steven Meyer calls the "dissociative writing of the middle period."[28] Approaching Stein through Mac Low thus deepens our historical image of her by illuminating elements of her work that have proved most influential for later experimentalists, not least of which are her reflections on the chanciness of language and history.

The aesthetic strategy that makes Stein's middle works so influential stems more from her interest in chance than from a "dissociative" impulse. For many readers, a book like *Tender Buttons* might at first seem detached from the actual because its oblique descriptions rarely yield to direct statements: it is "Claiming nothing, not claiming anything, not a claim in everything" (*TB*, 38). Some view this aversion to "claiming" as a way of putting the world at arm's length, securing a measure of autonomy for the literary work. Jennifer Ashton believes this gesture lends the text "a mathematical independence from experience as such."[29] The RAND book reminds us, however, that mathematics has its own messy relations with the actual and provides an unhappy figure for aesthetic autonomy. Others read Stein as more historically engaged. As Phoebe Davis claims, "Stein's literary experimentation is directly connected to her sense of history," including literary history and the historicity of the literary text.[30] Stein's middle works do not disengage from the historically actual world but rather contest a deterministic sense of history. As Kelly Wagers has argued, Stein mounts a "resistance to 'historical necessity'" by seeking to "contest and reform the narrative of necessary historical 'progress.'"[31] This opposition to historical necessity enables Stein to mediate between pragmatic, worldly interests and a composed attitude of aesthetic disinterest. Specifically, the forms of chance in her middle texts disrupt practical concern and sustain aesthetic, formal attentions. As I demonstrate below, *Tender Buttons* describes objects that happen to be lying around but does not ascribe meaning to their presence; when the shapes and sounds of its words exhibit formal order, they underscore the contingency of verbal form itself. As such work develops "an arrangement in a system to pointing," a presentational strategy, it does not set itself apart from the actual but seeks new ways of "pointing" out the chanciness of what happens, the turbulence of the world, as an occasion for aesthetic response, for poetic composition (*TB*, 11).

Some thirty-three years after *Tender Buttons*, Stein offered more explicit comments on the relation between contingency and value judgments. These appear in a short prose piece, "Reflection on the Atomic Bomb"

(1947), apparently her last writing before death, and they make it easier to discern an engagement with chance in her earlier work. Using a vocabulary of "interest," she describes how the contingency of nuclear war disrupts pragmatic ways of thinking and perhaps thereby opens a space for the aesthetic:

> They asked me what I thought of the atomic bomb. I said I had not been able to take any interest in it. . . . What is the use, if they really are as destructive as all that there is nothing left and if there is nothing there is nobody to be interested and nothing to be interested about. . . . Sure it will destroy a lot and kill a lot, but it's the living that are interesting not the way of killing them, because if there were not a lot left living how could there be any interest in destruction. . . . Everybody gets so much information all day long that they lose their common sense.[32]

Stein takes no "interest" in the bomb because nuclear cataclysm would unsettle normal forms of value. Instead, she advocates a localized "common sense" that sustains her observations of life's quotidian materials: "Everybody gets so much information all day long that they lose their common sense." We can safeguard practical interest by ignoring the torrent of geopolitical information that leads us to fear the bomb. In these early years of information technology, as Los Alamos scientists used computers to design nuclear weapons, Stein cautioned against the torrent of information electronic media could bring. As an isolationist gesture, her defense of local "interest" suggests a parochial conservatism discernible in much of her late writing, but in its dedication to the present scene and its hesitancy to ascribe value, the "Reflection" also recalls her earlier, more radical search for a new language of description. At first it reads as a note on the absurdity of total destruction, but this piece also begins to imagine a place for the aesthetic in a chaotic world of global information flows. In Stein's view, one can preserve "common sense" by avoiding information overload.[33] By contrast, too much information raises senseless anxieties about contingencies beyond one's control, including events that, if they happened, would radically disrupt human life. The "Reflection" suggests the better response to "so much information" is not anxiety but composure, and Stein's middle works anticipate this point. Through composed, formal attention, they explore the chanciness of language and history without arousing the anxieties that another avatar of chance, the atomic bomb, occasions for others.

By arguing that Stein maintains composure in relation to historical contingency, that her texts remain composed, I want to link her response to chance with her later theory of "composition," a term Stein uses to describe the historicity of aesthetic objects.[34] As Kristin Bergen notes, Stein uses "composition" to refer to "history itself," drawing a connection between aesthetic production and "the dominant mode of daily life for a given period." Especially in times of war, Bergen argues, this sense of "composition" enables Stein to render violent historical disorder "as a politically neutral formal structure or arrangement."[35] Expressions of anxiety emerge when Mac Low renders historical contingency through the procedural play of chance, but Stein carefully composes her descriptions of the contingent, of whatever happens, and thereby models a more placid, composed attitude about the chanciness of history.

Stein is far from alone in taking the play of random information as a gateway to aesthetic production. Stéphane Mallarmé comes to mind, as do later figures such as Marcel Duchamp, William S. Burroughs, John Cage, the OuLiPo, and indeed Mac Low. Even the scientists at Los Alamos can be seen in this light, since random numbers provide them an aesthetic figure for otherwise inscrutable subatomic processes. Stein's approach is notable, though, because it eschews the flimsy mysticism often deployed to justify chance-operational poetry. She instead gives a practical account of the relations among information, contingency, and poetic composition. As a theory of these relations, Stein's "Reflection" provides a guide for interpreting texts she wrote decades earlier. *Tender Buttons* evokes the disorder of a cluttered apartment without ascribing meaning to what appears where. Nor does any personal voice emerge to worry about the mess. Mac Low's rewritings, by contrast, produce the semblance of an expressive voice largely absent from Stein's originals. Writing after a Cold War that reified the possible end of history in the image of the nuclear bomb, Mac Low highlights the attitude of composure in Stein's work by contrasting it with the more anxious lines he renders from it.

The formal and descriptive techniques of Stein's middle texts enable her to develop this aesthetic response to chance. While writing *Tender Buttons,* Stein considered subtitling it "Studies in Description" (a phrase that did appear in marketing materials), and the book is known for renovating the language of description.[36] In this and related works, Stein curates an open attention to the contingency of whatever happens to be present, to what Roland Barthes calls "the vast disorder of objects."[37] At the end of *Tender*

Buttons, for instance, the mention of "a magnificent asparagus and also a fountain" might be seen as an intentional decision on Stein's part, but these details might just as plausibly stem from the possibility that Stein ate asparagus that day and not, say, broccoli or that she happened to write in sight of a fountain and not, say, a topiary. Her particular subjects seem accidental and secondary to the descriptive endeavor per se. Hence, when Juliana Spahr refers to *Tender Buttons* as "a possible description of a domestic space," the emphasis should be upon "possible" rather than "domestic"; the book teaches us that its particulars could easily be otherwise.[38] It is not enough to say, with the structurally minded readers of Stein, that it matters less *what* Stein describes than *how* she does. Rather, in their particularity, the things she describes remain central to the experience of reading but still seem haphazard, jumbled, not carefully chosen. If it seems superficial or derogatory to claim that, in their relative arbitrariness, Stein's descriptions primarily underscore the contingency of the present, then this impression should indicate the tenacity with which our "hermeneutic compulsion" clings to the meaningfulness of the artwork's details and recoils from contingency.

From a certain angle, any kind of description seems to arise from the play of chance. Description as such differs from selection, framing, arrangement, or figuration because it demands open attention to whatever appears. In this respect, it differs even from observation, with its scientific and forensic connotations. What would it mean to resist the revelatory gesture by which critics expose "mere description" as advancing an implicit judgment or showcasing a literary technique? It might mean lingering with how description submits itself to the facticity of appearances, the immutable happenstance of what is there. Here description functions as the degree-zero of appearance in writing, as the point where language shows a world without claiming anything about it, and the arbitrary play of verbal forms is its stylistics. Adorno views description in this way and suggests that it risks violating "an aesthetic imperative: the requirement that everything be required."[39] Normally, art "divests itself of its contingency" by imagining its particulars as formally "integral" to the whole.[40] In this sense, the artwork contests "not rationality" in general "but rationality's rigid opposition to the particular," to the accidental asparagus.[41] If the details of a description are "not as such construable out of any overarching *ordo*," not rationalized within a system, then whatever appears "always has an element of positedness, of game rules and contingency," of chance rather than necessity.[42] One might even have arranged the subject of description beforehand (as with the painter's still life), but rigorous description faithfully attends to the smallest accidents (the particular drape of a curtain, the blemish on a certain fruit). To understand description as a kind of chance operation helps explain why many postwar experimentalists interested in chance operations also sought new forms of description. Even in the hands of William Carlos Williams, Marianne Moore, or T. S. Eliot, description remains a relatively belle-lettristic affair, rewarding grace and virtuosic technique. Its later alignment with chance helped sustain the flatter, less mannered descriptive techniques that accompanied procedural methods in postwar poetry and beyond.

The play of chance in Stein's middle work is easiest to discern at the level of form. The sounds and shapes of her words ratify the arbitrariness of the sign, underscoring the contingent histories of literary genres and technologies. Like any prose poem, *Tender Buttons* troubles the distinction between prose and poetry, but *A Long Gay Book* and *Pink Melon Joy* more pointedly test this difference. Though formatted as prose, both contain series of short, one-line paragraphs that look like verse, and these draw a visual equivalence between a paragraph break and a line break, since a few short "paragraphs" in sequence give the appearance of verse. Even in the longer prose blocks of *Tender Buttons,* eye-rhymes insinuate a vertical sense of visual form more common in verse. In the first edition, Stein's repetitive syntax causes recurrent phrases to align vertically, making a visual pattern:

> A lamp is not the only sign of glass. The lamp and
> the cake are not the only sign of stone. The lamp and
> the cake and the cover are not the only necessity
> altogether.[43]

This formal effect has nothing to do with the meaning of the words or even the accident of their shapes, only with the way Stein's phrasing falls across the page. Later editions and collections have wider margins, so these eye-rhymes appear more sparsely and in different places. Even in the first edition, Stein cannot have decided where they would appear, since she did not oversee the book's production. The eye-rhymes are a formal effect of her syntactic choices, but their frequency and location are contingent on page width and other technical circumstances. In this sense, they emerge from a kind of chance operation that folds Stein's prose against itself at the arbitrary interval of the page width.

Other kinds of formal play persist across editions of *Tender Buttons,* and these too highlight the contingency of generic conventions and print technologies. The series of short "Chicken" passages in the second section, for example, follows the conventions for formatting prose, but as with the short paragraphs in *A Long Gay Book* and *Pink Melon Joy* mentioned above, the brevity of these sections makes them look like a spare poem:

CHICKEN.

Pheasant and chicken, chicken is a peculiar third.

CHICKEN.

Alas a dirty word, alas a dirty third alas a dirty third, alas
a dirty bird.

CHICKEN.

Alas a doubt in case of more go to say what it is cress.
What is it. Mean. Potatoe. Loaves.

CHICKEN.

Stick stick call then, stick stick sticking, sticking with a
chicken. Sticking in a extra succession, sticking in. (*TB,* 54–55)

The visual sparseness and patterning suggest verse, but these emerge from the same prose formatting conventions operative throughout the book. The passage's aural form, meanwhile, reads like a senseless incantation, while also suggesting a less arbitrary figural strategy: the repeated "alas" sounds like "Alice," Stein's partner, so Stein may call Alice "a dirty bird." Pamela Hadas reads *Tender Buttons* as a chronicle of Stein's alienation from her brother Leo in favor of Alice. "Alas a dirty word" makes her lover's name taboo, and "alas a dirty third" calls Alice an unhappy third party, disrupting the siblings' relationship. If Alice is a dirty bird, perhaps she is a "chicken," in which case Stein's "sticking with a chicken" means she will stick with Alice despite these tensions. Though the whole of *Tender Buttons* can be read as an epithalamion, the play on alas/Alice remains confined to this passage, reflecting Stein's preference for nonce figures. Even within this passage, Stein's language outpaces the ordering impulses of formal interpretation. Why mention "Pheasant"? Why "Loaves"? Why the archaic spelling of "Potatoe"? These words mean something more than how they look and sound, but this excess remains arbitrary and obscure. It emerges from the play of chance at multiple levels of Stein's writing, and it evades the capture of critical reading strategies.

Stein's formal play is local and occasional. It disrupts the generic and technological grounds for interpretation. The sound of a rhyme and the form of a visual pattern lack the sense of intention by which they might support a sustained reading strategy. The formal features of her work emerge through the contingency of its technological presentations, just as her descriptions are responsive to the chanciness of what appears. Together, these formal and descriptive strategies enable Stein to develop an aesthetic response to chance that provides, decades in advance, an alternative to the paranoid response to geopolitical contingencies that she derides in her "Reflection." Whereas Mac Low's chance-operational rewritings seem to voice anxiety about the histories in which they become involved, Stein takes chance as an occasion for a kind of poetic composure that imagines a "common sense" model of value, of "interest," beyond the threatening chanciness of geopolitical information.

Some interpretations of Stein elide her interest in chance for reasons having to do as much with interpretation in general as with Stein in particular. Philosophers both ancient and modern have seen chance as disrupting physical determinism and making agentive intention possible, including the intention to signify, but theorists of interpretation have not explored this interface between chance and intention very successfully.[44] Texts obviously express meanings their authors intend, but an array of accidents also informs any textual field, from the arbitrary shapes and sounds of a language to the width of a page, the pitch of a voice, or the historical contexts of writing. Any of these accidents may complicate the recovery of an author's meaning, but more important for us, such accidents also ground her ability to intend a meaning in the first place. Not only is it often impossible to say whether a textual detail is intended or accidental, but also writers often embrace this unsteady horizon of accidental meaning as a structure for literary play. Indeed, the play of chance operates as a mode of lyric exemption in precisely this way: it enables poetry to trouble the distinction between intended signification and mere accident. *Either* to divorce Stein's texts from her authorial intentions *or* to claim she intentionally shapes every meaningful facet of them is to ignore her interest in chance as a basis for aesthetic production and a threat to normal discourses of value. Stein makes these interpretive challenges more visible than Mac

Low because it is easy to marginalize chance's role in procedural poetics by emphasizing the structure of the procedure, as Mac Low and I do above.

The descriptive and formal arbitrariness of Stein's middle texts has registered in the criticism as well. Interpretations of her most difficult work have become so diverse that the sole trait everyone can recognize in a Stein text might be its ability to accommodate a wide range of responses. Marjorie Perloff begins her influential reading of *Tender Buttons* by calling it "particularly resistant to interpretation," but she then cites several different interpretations of the book and offers one of her own.[45] In a more recent commentary, Spahr ably captures the book's capacity to host divergent interpretations:

> It has been said that it presents "a woman-centered, revisionary spirituality." That there is "increasingly explicit" dildo imagery, and also that it is about oral sex with a circumcised penis, and then there is a clitoris that is rubbed with a rubber cock too. Others argue it is about a woman's nipples. No, still others reply, it is about the early buddings of a plant.[46]

The diversity of interpretations is striking. Perloff's reading has endured partly because she recognizes the book's malleability in the hands of such readers, its power to "manifest the arbitrariness of its discourse" and thereby accommodate varying responses. She sees the details in *Tender Buttons* as "false leads" soliciting hermeneutic efforts that it then embarrasses with its arbitrariness.[47] The problem is not that nothing sure can be said about *Tender Buttons* and related works; it's that altogether too much can be said about them. Insomuch as this malleability provides unusual leeway for scholars to recruit these texts for their own ends, it also enables Stein to negotiate her work's exposure to historical contingencies, including the contingent range of readings it might occasion in the future. My own reading of Stein necessarily operates within the same sense of contingency, and if it is effective, it clarifies the role of chance and arbitrariness in Stein's thinking about the historicity and value of her work.

Close readings of *Tender Buttons* do not resolve the play of chance across the text, for the selection of evidence seems inevitably motivated by the argumentative ends of the critic. Sianne Ngai, for instance, provides good evidence that cuteness is the overriding aesthetic of *Tender Buttons*. She defines the cute object as having "a softness that invites physical touching—or, to use a more provocative verb, fondling."[48] Yet the book also deploys edgier, less comfortable sexual imagery. The short section titled "PEELED PENCIL, CHOKE" simply reads, "Rub her coke" (*TB*, 31). A peeled pencil suggests an exposed penis, and "choke" sounds sadomasochistic. The following line, to which Spahr alludes above, plays on "rub her cock" and "rubber cock"—images more queer than cuddly. The eroticism of "RED ROSES" proves equally gritty: "A cool red rose and a pink cut pink, a collapse and a sold hole, a little less hot" (*TB*, 26). Both the section title and the "pink cut" suggest a vulva, which then becomes "a sold hole," suggesting prostitution. Only by focusing on "less hot" moments does Ngai cast *Tender Buttons* as cute rather than kinky; her reading seems both correct and inevitably partial, selective. Ngai situates cuteness among a set of minor aesthetic categories by which artworks "bear witness to their contingency." The cute object's pliancy makes it durable: "smallness and blobishness suggests greater malleability" and thus a "capacity to withstand rough handling."[49] This malleability enables cute objects to absorb and recover from antagonisms of many kinds. Such a strategy of pliant resiliency, Ngai argues, helps a variety of avant-gardists to bring their work in relation to a sense of history as threateningly violent. Alongside her readings of Stein, she discusses the famously cute work of Japanese artists Yoshitomo Nara and Takashi Murakami, who often place cute figures in "a sinister or implicitly menacing environment." Ngai sees Murakami's cartoon mushrooms as "recalling the mushroom clouds of the atomic bombs" dropped on Japan.[50] Instead of strongly defensive responses to such violence, which would dramatize the artwork's powerlessness, a malleable cuteness helps artworks absorb the violent impacts of historical contingencies without becoming fatally damaged or undone. In the same way, the malleability of Stein's work helps it to cope with the disruptiveness of being historical: instead of withering at the prospect of historicity, her work absorbs and reflects the sense of contingency that is the fundamental mark of the historical.

As qualities influential in the reception of Stein's middle works, their malleability and descriptive arbitrariness help to sustain a strategy of historical regard, a way of responding to historical uncertainty, rather than a means to destabilize the reader's experience of a text or the writer's production of it. My emphasis on the arbitrariness and interpretive pliancy of *Tender Buttons* may thus recall Perloff's notion of a "poetics of indeterminacy," but our differences matter. Perloff calls *Tender Buttons* a "Dada joke," confirming the lines of inheritance Mac Low's work has since invited us to question, and she frames the book's arbitrary descriptions as experiments with "naming," with the arbitrariness of the sign, instead of linking them with broader discourses about contingency—historical, aesthetic, and philosophical.[51] Instead of cementing a literary-historical lineage or activating an empty play of signs, the contingency and flexibility of Stein's middle works enable them to negotiate the instabilities of historical relations and the apparent chanciness of the present.

MAKING MEANING FROM A MILLION RANDOM DIGITS

Having outlined Mac Low's methods and examined the source texts by Stein, we can learn more about their shared conceptual terrain by focusing on the equipment that links them: the RAND book. Production of *A Million Random Digits* began at Los Alamos in the spring of 1947, and the digits were tested and used internally between then and their 1955 publication by the Free Press. Mac Low would discover the book by 1958. Though he used it in early works such as "Sade Suit" (1959) and intermittently thereafter, his most extensive use of it was for the *Stein* series.[52] The scientists for whom the book was made used it for "various kinds of experimental probability procedures," the most common of which were the "random walk" methods developed at Los Alamos and nicknamed "the Monte Carlo method."[53] Invented by mathematician Stanislaw Ulam, Monte Carlo mathematics uses random sampling to understand the behavior of complex systems such as financial markets or turbulent fluid environments. Such systems involve so many interdependent variables that they seem chaotic, so random sampling proves effective for modeling them. For example, Monte Carlo sampling would enable me to compare the areas of two irregular shapes drawn on the ground. If I scattered grains of rice across these shapes at roughly even density, I could then count the grains that fell inside each shape. The totals would provide a ratio comparing the areas of the two shapes—where more densely scattered rice yields a more precise ratio but a more arduous calculation.

Such methods developed in tandem with atomic physics. Already in the 1930s, Enrico Fermi had found random sampling helpful for calculating neutron diffusion, but these techniques did not see routine use until Ulam shared his ideas with others at Los Alamos in the late 1940s.[54] They became crucial for modeling the behavior of atoms during a nuclear reaction and thus facilitated the design of thermonuclear weapons. The "random walk" version of this technique, by which one inserts random values at certain steps in a calculation, became famous thanks to economist Burton Malkiel's

1973 bestseller *A Random Walk down Wall Street*. The scientists who developed it in the 1940s used it to trace "a genealogical history of an individual neutron" in a simulated nuclear reaction.[55] As a means of searching out virtual origins and tracing material pathways, then, the book helped its users negotiate the interface between chance and history from the very start. The usefulness of such methods at Los Alamos "led to a desire for a large supply of random digits, of sufficiently high quality so that the user wouldn't have to question whether they were good enough for his particular application."[56] To answer this need, the RAND Corporation set about producing the tables that would be published as *A Million Random Digits with 100,000 Normal Deviates*. A newsletter at Los Alamos joked that librarians would file the book under "abnormal psychology."[57] Figure 5 depicts a randomly selected, scaled-down facsimile of roughly one-quarter page.

34					TABLE OF RANDOM DIGITS	
01650	94988	12022	77021	60277	39048	03087
01651	72363	40974	09594	10276	09631	43203
01652	74967	66480	83894	82989	24784	42757
01653	26236	32399	81419	47377	93952	89101
01654	05632	68465	67842	85597	02094	42059
01655	67352	41392	17545	30949	87565	83820
01656	92727	35027	03117	80848	74559	96797
01657	18223	91136	39695	39943	77413	48937
01658	80723	91394	02992	11530	67845	05881
01659	75007	85671	88211	55080	15581	02685
01660	60050	80463	30926	74970	38951	14928
01661	69715	28522	73974	99491	50647	20252
01662	92096	43555	48882	60717	07963	39375
01663	35482	63353	08086	66635	71009	95777
01664	24879	78061	38949	21123	28430	72627
01665	68985	60486	58133	07709	25899	68531
01666	96601	96785	20850	70389	74637	34020
01667	66706	67664	93292	05934	71050	68192
01668	39273	41912	40198	36441	89472	38835
01669	42539	21771	58672	71421	92528	67229

Figure 5. This portion of a randomly selected page from the RAND book shows page and line numbers at left. Photograph by the author.

The physical gesture of opening a book carries with it a deeply entrenched anticipation of meaning, so one half expects to find something meaningful as one opens *A Million Random Digits*. But the book contains a true nonsense of digits. To generate them, a random-frequency pulse was coupled to an electronic "roulette wheel" bearing the numbers 0 through 9. Imagine spinning such a wheel at 3,000 revolutions per second and allowing a random pulse to dictate where it would stop. This is how the digits were generated, but some uncertainty remains about the source of the random pulse. Given the nuclear context, some believe it came from a Geiger counter aimed at a piece of uranium, which decays at a steady rate but discharges particles at random intervals (hence the Geiger counter's chaotic clicking). In his history of RAND, however, Willis H. Ware suggests that the random pulse, "given the technology of the time, would have been a vacuum-tube machine," which traces the excitation of electrons in a magnetized diode.[58] Either way, the random digits were generated using the same random subatomic motion they served to model. To avoid typographical error, the book was printed by photo-offset from the original tables, and the page and line numbers remind us that the digits have since remained totally fixed. Just as Mac Low's procedure works between chance and determinist rules, so does the disorder of these digits remain rigorously stable, unchanged since the first printing, yet not for that reason any more legible or meaningful.

Two organizations gave rise to *A Million Random Digits*: the RAND Corporation and the Los Alamos Scientific Laboratory, as it was then called. Their activities in the 1940s and 1950s were typical of a growing network of institutions, collaborations among the military, universities, and private industry that pursued basic and applied research for the many electronic innovations beginning to emerge in the postwar years. Los Alamos was first designated "Site Y" as part of the Manhattan Project. It was managed by the University of California, but until the end of the war, only one UC official, treasurer Robert Underhill, knew its purpose or even which state it was in.[59] The RAND Corporation, whose name stands for "Research ANd Development," emerged from a collaboration between Douglas Aircraft Company and the U.S. Air Force, and its first product was the prescient *Preliminary Design of an Experimental World-Circling Spaceship* (1946). Similar organizations included the Lawrence Livermore National Laboratory, another University of California affiliate focused on nuclear weapons; the Institute for Advanced Study, research home of Albert Einstein,

John von Neumann, and others; and AT&T's Bell Labs, where scientists invented the solid-state transistor the same year work on the RAND book began. Decades before Google and other Silicon Valley firms became known for their casual work environments, these groups recognized that a little disorder in the workplace fosters innovation better than strict corporate protocols. Researchers were rarely held to specific performance standards and were often encouraged to collaborate, cross disciplinary boundaries, and follow lines of inquiry without immediate practical goals in mind. Many of the greatest advances in these labs stemmed from a serendipitous collaboration or a seemingly pointless diversion.[60]

The same flair for disorder often shaped research topics themselves. Though much of this work would later be called "information science," the labs' most successful researchers often sought ways to grapple with whatever opposed information, logic, and order—including phenomena like randomness, uncertainty, noise, and distortion. In this sense, "information technology" may be a misnomer, for these devices emerged from scientific efforts to theorize and manage the others of information. For instance, Norbert Weiner is remembered as the father of cybernetics, the science of command and control, but humanities scholars have less to say about his seminal contributions to stochastic mathematics or about feedback, the guiding concept of cybernetics, which musicians know can be powerfully disruptive. Claude Shannon, working at Bell Labs, invented information theory in his 1948 article "A Mathematical Theory of Communication," but from its opening paragraph, the article is as concerned with entropy and "the effect of noise in the channel" as with information per se.[61] A primary innovation of Shannon's essay is his use of Markov chains to produce a formal, binary measure of entropy in communications; he quantizes information only by first quantizing its other, entropy. Shannon's diagram of a communication system looks like something out of Saussure, except that it includes a "noise source" intervening between sender and receiver (Figure 6). Why, then, should Saussure be remembered for the arbitrariness of the sign while Shannon and his cohort are seen as informationalists? To use the language of this technical scene from which the RAND book emerged, I have argued that Stein and Mac Low are more interested in the poetics of the noisy channel than the informatics of the message.[62] The same might be said of a wide array of poets who care more about the "noise" of language than its informational character.

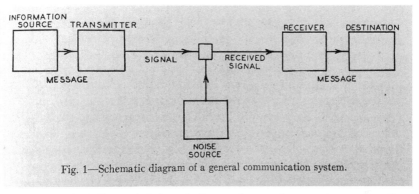

Fig. 1—Schematic diagram of a general communication system.

Figure 6. In Claude Shannon's "schematic diagram of a general communication system," the central, unmarked node is the injunction of noise into the channel. Reprinted with permission of Nokia Corporation.

Shannon's engagement with randomness in "A Mathematical Theory of Communication" leads him to perform a writing experiment that sets a remarkably close precedent for Mac Low's. Shannon imagines a series of "simple artificial languages" in order to quantify the information each contains (387). He starts with a twenty-seven-unit vocabulary comprising the twenty-six letters and a space. His first language consists simply of letters (and spaces) selected at random, producing gibberish. The second weighs the probability of selecting each letter according to its frequency in English, making *e* appear more often than *d*, for instance. The third language introduces a "digram structure," such that "after a letter is chosen, the next one is chosen in accordance with the frequencies with which the various letters follow the first one." This language remains nonsensical, but familiar formations such as "ch," "re," and "ea" begin to appear. Here is part of the sample Shannon gives: "ON IE ANTSOUTINYS ARE T INCTORE" (388). Shannon then imagines a fourth language that reflects a "trigram structure" in spelling; it considers the two previous letters when making a selection. His next artificial language works in "word units," selecting words with frequencies appropriate to natural English. His sixth and final artificial language uses "Second-Order Word Approximation," in which "word transition probabilities are correct," meaning that each word's selection probability is weighted in terms of its likeliness to appear after the previously selected word. Shannon provides this sample of second-order word approximation:

THE HEAD AND IN FRONTAL ATTACK ON AN ENGLISH WRITER
THAT THE CHARACTER OF THIS POINT IS THEREFORE
ANOTHER METHOD FOR THE LETTERS THAT THE TIME OF
WHO EVER TOLD THE PROBLEM FOR AN UNEXPECTED (388)

The similarities between Shannon's method and Mac Low's are quite
striking. Shannon uses "a book of random numbers" to make selections—
though, in 1948, it cannot have been the RAND book (389). Later in the
procedure, he "opens a book at random," reads until a desired letter or
word is encountered, and records the next letter or word in the text. One
might reasonably call Shannon's process a kind of reading-through text
selection, to use Mac Low's terminology.

Shannon interprets the results of his experiments in much the same
way as I have begun to interpret the *Stein* poems. Like Mac Low's work,
the quotation above seems very close to syntactic and semantic coherence,
but perhaps not quite there. As Shannon assesses his artificial languages,
he notes their increasing "resemblance to ordinary English" (388). He
claims that the texts "have reasonably good structure out to about twice
the range that is taken into account in their construction." In other words,
"four-letter sequences" from the diagram sample "can usually be fitted into
good sentences," and "sequences of four or more words" in the final sample "can easily be placed in sentences without unusual or strained constructions" (388–89). He specifically notes the legibility of the ten-word
sequence "attack on an English writer that the character of this," and he
normalizes capitalization in this manner when he cites it (389). Shannon's
response to these stochastically generated artificial languages thus anticipates my own response to Mac Low's work. Shannon knows very well that
the strings of words express nothing more than an interplay between randomness and statistical weights, yet he hears meaning begin to emerge from
these chance-operational procedures. He even notes that "it would be interesting" to produce higher-order selections, "but the labor involved becomes
enormous" when two-word probabilities come into play. Shannon's most
direct aim in his article is not to observe the emergent legibility of stochastically constructed word strings. Rather, he produces these strings to illustrate his quantitative measurement of how much information a message
contains. Nonetheless, one wonders whether he would find further experiments of this sort "interesting," for as Mac Low's work demonstrates,
these strings of words seem increasingly expressive.

After the publication of his article on information theory, Shannon decided to research the computability of chess. This might seem an eccentric turn for the employee of a telephone company, but scientists and poets alike have long recognized the links among games, mathematics, and the kinds of disorder that interested Shannon. Chance methods such as Mac Low's sustain a playful unpredictability in the creative process. They frame poetic production as a game, as rule-bound yet chancy, and in this sense they invert the scientific response to games of chance. On one hand, a poet might view chance as an end in itself. Mac Low acknowledges his equipment's links to gaming, as when he notes that "random-digit tables are really computerized roulette wheels" ("Poetics," 171). When asked how his pieces compare to games, he suggests a pacifist ethic of cooperation: unlike most games, "these games I call pieces are noncompetitive" ("Poetics," 184). As we shall see, Mac Low viewed chance as a reparative force, enabling him not only to unshoulder the "burden of 'expression'" but also to cleanse a text of unhappy historical contexts. He takes the play of chance as an end in itself. On the other hand, weapons designers using the RAND book began with a purposeless, aleatoric game and instrumentalized it to destructive ends. The idea for the Monte Carlo method purportedly came to Ulam as he played solitaire. He realized that the easiest way to calculate the probability of a solitaire game's coming to completion would be to lay out several sample games and record the number of moves each allowed, extrapolating a probability from these random samples.[63] This new technique was named "the Monte Carlo method" in honor of Ulam's uncle, who liked to gamble. Earlier, scientists had playfully referred to the first nuclear bomb as "the gadget," and of course, the Monte Carlo method would contribute to a decades-long war game.

The association between atomic weapons and games had become a cultural commonplace by 1950, when the U.S. Armed Forces released a short documentary called *The Basic Physics of an Atomic Bomb*. After explaining the bomb, the narrator reflects, "In building our bomb, we've been playing a game, toying in a broad general way with theories already worked out in detail and executed by those who, during World War II, developed and produced the most destructive weapons of all time."[64] During the subsequent Cold War, the contingency of nuclear catastrophe was figured as a game with sufficient frequency that Derrida, theorizing bad luck in 1982, could coherently ask, "What are the chances of my losing at a game or for the neutron bomb to be dropped?" These had begun to seem equivalent: both contingencies bring us "into contact with what *falls*"—a bomb from the sky, dice from a hand, atoms through the void—all according to the play of chance.[65] The RAND book thus reminds us that games do not just provide a convenient analogy for military activities, nor can we always view an aesthetic of randomness as a harmless amusement, contained by its own field of play.[66] Rather, games of chance can produce military violence and imperial projects in the first instance. By highlighting the real productive powers of chance, the RAND book situates it not as a mere game, but as a kind of interface, a mode of contact between the world and the systems by which we model, represent, produce, and potentially destroy it.

Through the *Stein* series, Mac Low works to recuperate *A Million Random Digits*, putting it to less destructive use. In the process, however, the RAND book also involves Mac Low's work in a discourse about nuclear war and, more broadly, opens questions about how historicist interpretation might address poems such as these. Just as Mac Low's procedure combines chance operations with deterministic rules, the RAND book intertwines randomness with a more causalist view of political motives and consequences. It solicits a sensitivity to randomness far beyond the specific origin and uses of its digits, for the book helped produce a threat of nuclear apocalypse that transformed historiological thinking *tout court*. Under the threat of thermonuclear war, both history and its potential closure became perilously bound up with the instabilities (physical, ethical, political) of a specific technology.[67] Historical interpretation of the *Stein* series in the context of its compositional equipment thus operates through an idea of history as itself shaped by such instabilities. In other words, when random decisions made in a poetic game or a war game can inform the outcome of a narrative, we lose the taut sense of causation that stabilizes historical meanings. If things just happen by chance and have indeterminate effects on what follows, what knowledge can we gain from studying the past? An effort to read the *Stein* series in its historical contexts will not only conjure Cold War anxieties about the end of history but also bring into focus chance's broader tendency to complicate historical sense-making.

Although Mac Low used chance operations to produce the *Stein* series, the poems often seem to express anxieties not evident in Stein's originals. Where Stein takes disorder and contingency as occasions for aesthetic play, Mac Low's rewritings imbue their reader with anxiety about the randomness through which the poems are generated. They prompt readers to worry that the randomness of what happens will unsettle our frames for ascribing value, including the literary value we attribute to these texts themselves. This more anxious mood finds expression in the opening stanzas of "Who Is Showing Us What Happened in This Corner? (Stein 103)":

> Who is showing us this unordered spectacle?
> They are recklessly making exchanges.
>
> Not likely in the closet, she connects singing with a message.
> He does not like to show himself singing in the closet there.
> Hanging in it sooner, for no side is established there.
>
> Suppose her cloak is spread.
> A single cane is attractive.
> When a white stamp stops being shown, it becomes increasingly fitting.
> Housing and shows preserve their patients.
> Pack some lead-colored glasses.
>
> No stone loses the inclination to be shown.
> Isn't it the same as any sign at a spectacle?
> No spectacle makes a season.
> She strokes wood on the spot as a charm.
> Between stomachs the lightening lace is reckless and did less.
> No dog needs to be wearing lace at the wrist in the summer.
> Doesn't the nearest spot show?[68]

This poem's source text, *Tender Buttons*, contains not a single question mark, but "Stein 103" has twelve, meaning the punctuation is Mac Low's. Here it conveys a sense of anxiety. Of course, Stein's book contains plenty of questions: "Nickel, what is nickel," for instance, or "Does this change" (*TB*, 11). But without question marks, these read more as assertions of a question than solicitations of an answer. In the rewriting too, syntax alone suffices to mark the questions as such; the question marks express a diacritical anxiety not evident in the source text.[69]

Reading Mac Low, I often describe "anxiety" as the referent of his poetic expression or the occasion for a defensive response. Here anxiety refers not to the Freudian family drama Harold Bloom invokes in his account of poetic revision. Rather, it refers to the midcentury "age of anxiety" during which, as William Faulkner put it in his 1950 Nobel Prize address: "There are no longer questions of the spirit. There is only the question: When will I be blown up?"[70] Instead of despecifying a condition of influence, this anxiety marks a historically specific way of thinking about chance and political violence. W. H. Auden might easily have been talking about the coming Cold War in *The Age of Anxiety* (1947), where he describes an era "when everybody is reduced to the anxious status of a shady character or a displaced person, when even the most prudent become worshipers of chance, and when, in comparison to the universal disorder of the world outside, his Bohemia seems as cosy and respectable as a suburban villa."[71] The chaos of war provides a potent figure for this historical anxiety, but it encompasses a much broader set of concerns about the contingent historical situation of a person or text and about how to respond to that contingency.

The *Stein* poems lead their reader to worry about disorder and contingency. Because they were produced through chance operations, they embody the same anxieties they occasion. Already in the first line, the "not unordered" description that opens *Tender Buttons* has become "this unordered spectacle" (*TB*, 11). Stein's double negative is calm and composed; her book is "not unordered" in the same way my desk is not a total mess. Mac Low's procedure yields a more overt disorder, yet one that seems capable of referring to itself. The opening line articulates a "who?" that raises concerns about showing and being shown, about the agency behind Mac Low's "reckless exchanges" with Stein. The poem leaves close readers with anxiety about the arbitrariness of its discourse. In contrast with Stein's aestheticizing response to chance, both Mac Low and his reader tend to worry about this "reckless" revision of her texts, and, indeed, to worry about worrying too much. Anxieties redouble as the poem conjugates "show" through its various tenses. In the first two stanzas, "show" is an active, transitive verb: one shows an audience ("us") a spectacle, or one shows oneself. In the third stanza, however, the past participle and plural noun suggest a more passive visibility. This passivity is fully expressed in the fourth, where "No stone loses its inclination to be shown." Of course, no stone has any inclination, any volition, at all. A stone on an incline has no choice but to roll, or to be shown, much as Stein cannot keep Mac Low from making an "unordered spectacle" of her words. The *Stein* poems express anxieties about the same randomness through which Mac Low produces them, leading us to ask how such randomness shapes the literary and political histories that stand between Stein and Mac Low, through which we think their continuities. This anxious mood may strike the reader most strongly in the command that closes the third stanza: "Pack some lead-colored glasses." The sufficiently paranoid reader will see these not as drinking glasses but as lead-colored spectacles, eyeglasses. And what would leaden eyeglasses provide if not a shield from radiation? If we had them, perhaps we could view Mac Low's poems without anxiety about the chance operations that threaten to make any reading of the *Stein* series seem arbitrary: they might help us toward the less anxious attitudes Stein models in the "Reflection" and elsewhere. As it stands, though, the *Stein* poems appear to express anxiety about the chancy conditions through which they emerge. These lines worry about the "unordered spectacle" they make of Stein's texts, leaving it for the reader to worry about which textual details are probative, which accidental, and to ask which literary and political developments lead from Stein's composed view of historical uncertainty to Mac Low's anxious response.

This is a challenging question about historical transit, about the lineages of experimental writing that link Stein with Mac Low. As I suggested above, the answer might first route Mac Low's chance operations through such European avant-gardists as Mallarmé, Duchamp, Georges Perec, Raymond Queneau, and André Breton. Indeed, the expatriate Gertrude Stein looms large in standard narratives about the American absorption of European experimentalism. The equipment mediating her exchange with Mac Low, however, is the product of a specifically U.S. military-industrial history. The search for an explicitly U.S. lineage of chance writing might upset the normal ways in which a literary work becomes marked as national and in which nationality informs constructions of literary history. There are hints of proceduralism in some of Stein's work—the exhaustive categorization of *The Making of Americans,* for instance—but she does not employ the explicitly procedural or stochastic techniques of Mac Low, the Fluxus group, the OuLiPo, or the conceptual writers.[72] Stein declared that "America is my country and Paris is my home town," but she does not provide an anchor for U.S. poets' use of chance operations per se.[73] The RAND book instead suggests less comfortable forms of poetic inheritance and literary nationality. If a poem is composed by a random process, perhaps its system of literary influences registers only contingently or arbitrarily within the text itself instead of concretizing chance techniques as a necessarily French (or, say, Brazilian or African American) dispensation. And if chance writing's national marking can arrive through a military reference volume, then such poetry might signify a violent sense of the national instead of expressing a happy cosmopolitanism. Many postwar and contemporary experimentalists, Mac Low included, see their work as sociopolitically liberative, anti-imperial, and pacifist, but through the *Stein* poems, the equipment of U.S. nuclear armament and capitalist imperialism intrudes upon this tradition, clashing with its political commitments. Still, *A Million Random Digits* provides only random digits, so perhaps it does not mark its intertexts with its own history. Its numerical disorder seems capable of disrupting the familiar causal terms of historical thinking, but as equipment for poetry, the book continually insinuates the militarism of its national significance.

Without strongly refuting the story of transatlantic influence, then, one might also locate the *Stein* series within a lineage of U.S. poetry about nuclear armament and Cold War anxiety or within a larger U.S. tradition of taking the atomic as a cipher for national identity. As early as the third line of Whitman's 1855 "Song of Myself," we read that "every atom belonging to me, as good belongs to you."[74] This is no passing reference: Mark Noble has traced Whitman's fascination with atomic science, arguing that the atomic embodies for Whitman "the pure democracy and thus broad possibility that are both inherent in and constitutive of nature," and of course, key to the U.S. political imaginary from Whitman's day to our own.[75] Mina Loy affiliates Stein with atomic science as early as 1924, calling her "Curie / . . . of vocabulary."[76] Stein welcomed such comparisons, saying in *Everybody's Autobiography* that "Einstein was the creative philosophic mind of the century and I have been the creative literary mind of the century."[77] U.S. poets' interest in the atomic remained strong into the 1950s, when we encounter Allen Ginsberg "listening to the crack of doom on the hydrogen jukebox."[78] Between then and now, many poets would continue to express an anxiety John Berryman called "the night sweats & the day sweats" over the "radioactive" global conflict.[79] In 1973, Robert Hass further involved Stein with anxieties about nuclear war by titling an edition of her uncollected writings *Reflection on the Atomic Bomb,* after the short essay discussed above. Even as we understand Mac Low's procedural reworking of Stein as part of a transatlantic lineage, we must also attend to the ways his unusual equipment makes visible less obvious lines of influence and historical motifs in U.S. poetry.[80]

Through *A Million Random Digits,* the *Stein* series also rehearses an ancient entwinement of poetry, randomness, and atomic science, making these twentieth-century contexts seem at once overdetermined and arbitrary. This entwinement goes back at least to the philosopher Lucretius, whose long poem *De rerum natura* (ca. 50 BCE) brought atomism to a Roman audience. For Lucretius, the chaotic motion of atoms in the void indicates a fundamental principle of randomness that he calls the *clinamen*, and this principle helps him explain everything from basic mechanics to free will. Many have noted similarities between the Lucretian *clinamen* and certain phenomena of modern physics, such as radioactive decay and Brownian motion. Like the *clinamen*, these suggest an aesthetic of irreducible physical randomness, the same aesthetic that made the RAND book scientifically useful. The *clinamen* also enables Lucretius to say why the universe exists at all. With "no clash / No blow" between swerving atoms, he writes, "nature" would have "begotten naught at all."[81] In modern discussions, the *clinamen*'s influence stems not from its ambivalent relation to physical causation but from its equally ambivalent relation to personal agency, though the two are in fact related. In a truly deterministic universe, Lucretius writes, "the mind / Should feel within a stern necessity" and free will would be at best an illusion.[82] But if physical laws entail randomness, free agency can emerge in the interstices between cause and effect opened by this play of chance.[83] Contrary to the assumption that chance-operational writing pursues the hygiene of an anti-egoic proceduralism, then, the discourse of the *clinamen* suggests that randomness provides the very kernel of free choice. When Mac Low disavows chance's role in his procedure, emphasizing the combination of determinism and "personal" decisions, he does not consider that his surprisingly expressive poems and, perhaps, his capacity for "personal" choice itself may be grounded in randomness.

The RAND book thus begins to endow its intertexts with its own destabilizing energies. As the above attempt at a nonce genealogy of "atomic" poetry suggests, the production of historical meaning often privileges causation and continuity, marginalizing accident and coincidence. If poets perform chance operations because such techniques enable an "escape from the confines of bias and personal preference," as Alison James writes, then it becomes difficult to describe their shared method as a literary tradition, since what they share is an interest in detaching their work from the contexts of received tradition.[84] Conversely, when procedural writing involves equipment as fraught as *A Million Random Digits*, the resulting texts become involved in other histories—military, technological, political—by way of the very equipment that had seemed to disrupt contextual relations.

Making historical sense of the *Stein* series thus heightens the anxieties these texts occasion, and this approach both energizes the anxious discourse of nuclear war and opens difficult questions about how to contextualize chance-operational poems. A less confounding approach to the relation between Stein and Mac Low might eschew these uncertainties about the histories from which Mac Low's work emerges, preferring instead to discuss the *Stein* series as an image of Stein's own futurity. Although experimental writers have claimed Stein as an influence since at least the 1970s, recent historicist criticism has reemphasized her political and aesthetic conservatism, questioning her status as an avant-garde saint. When the *Stein* poems render an image of Stein's work in the future of its reception, they operate as *readings* of Stein, readings that clarify the stakes of her influence upon recent experimental poetry by providing a counterpoint to her more composed, pragmatic responses to disorder and uncertainty.

READING EXPERIMENTS

As readings, the *Stein* poems do not contain critical discourse in the normative sense. Far from it. But they do respond to and change our view of their source materials, as any good reading does. One finds reasons to approach the *Stein* poems as readings not only in Mac Low's own comments but also in the broader commitments of the poetic lineage to which he belongs. Mac Low and many others have cast Stein as a vital precursor to more recent experimental writing, so his decision to rewrite her may reflect a devoted readership. In 1978, he contributed to a feature called "Reading Stein" for the journal *L=A=N=G=U=A=G=E,* in which several writers commented on an excerpt from *Tender Buttons.* There Mac Low describes how "hearing the sounds" of Stein's text leads him to "go from word to word" instead of reading synthetically, a process echoed in the word-by-word counting of his procedure.[85] As we have seen, he refers to the procedure itself as "diastic reading" and "reading-through text-selection," as if to equate textual manipulation with interpretation. Bernstein sees "language speaking for itself, restored to its autonomy," in Mac Low's work, but "these ideas are about styles of reading not writing."[86] Similarly, Steve McCaffery argues that Mac Low's procedural writing "fails to yield much aesthetic gratification," so he believes its "contribution is best situated inside the cultural history of reading . . . and not in the history of writing that we term 'literature.'"[87] Though the *Stein* poems do not explicitly advance critical claims about Stein, the rhetoric of reading resignifies Mac

Low's engagement with his predecessor. Such rhetoric is more than a poetic conceit; it opens the *Stein* poems onto their literary context, positioning the series as an index of Stein's reputation and linking it with a long line of experimental poems that function as readings.

For Mac Low and other experimentalists, including the Language writers, who view him as both a peer and a guru, approaches to composition modeled on reading make it possible to resist traditional modes of poetic expression. These approaches often replace intimate poetic voicing with broader efforts to record a sociohistorical state of affairs. As Jed Rasula puts it, "to be a reader is to be the willing receptor of transformative agencies destined to either alter or confirm one's position in a social circuitry."[88] A poem that "reads" does not pursue individual invention or personal expression but reflects a social order. If it bears a specific relation to its author, it does not disclose her subjective interior but indicates her "*positionality*" within the total, historical, social fact."[89] As noted above, Mac Low welcomes such escapes from the "burden of expression" and believes chance operations help to efface the writer's "ego." When a poem no longer pursues expression, it may serve a mnemonic function, tracing the linguistic textures of the social reality from which it emerges. As Ron Silliman argues, "the work of each poet, each poem, is a response to a determinate coordinate of language and history."[90] Indeed, the *Stein* poems accomplish nothing if not a linguistic coordination of literary history, the one that made Stein an obvious choice for Mac Low's rewritings. Reading the *Stein* series in relation to Mac Low's procedure evokes other histories, among them the long history of chance-operational poetic techniques and the material history of the RAND book. Nevertheless, we have seen that the *Stein* poems seem to express anxiety about the same historical relations they evoke. They thereby suggest, first, that randomness not only disrupts historical relations but also informs them and, second, that chance operations and related procedures do not necessarily cleanse a text of expressiveness so thoroughly as Mac Low and his fellow experimentalists sometimes assume.

Those who imagine poetry beyond the conventions of lyric expression often emphasize the informational dimensions of poetic language. They thereby suggest that the afterlives of the lyric hinge upon the question of whether a poem conveys knowledge. For Rasula and Silliman, a poem that eschews the intimacies of personal voicing can instead give a picture of "the total, historical, social fact." The implication is that poems most expressive of personal interiority convey something other than "fact." This formula captures the information fetish of much anti-expressive writing, from late-1970s Language work to today's conceptualists and digital bricoleurs. It also clarifies the commitments of some groups who renovate the personal voice instead of rejecting it: deep image and late confessional poets, for instance, as well as many feminist and minority poets of social justice. Lost in the midst of this opposition, however, is a more dialectical understanding of poetic expression under which the very inwardness of the lyric voice indexes the sociopolitical world that produces it. As Silliman's phrasing suggests, anti-expressive writers yearn for a kind of poetic world picture, a global perspective whose unavailability becomes evident through the play of chance across multiple systems—scientific, historical, and indeed poetic. In Mac Low's case, lyric expression emerges from the very procedures expected to undercut it. Because his texts seem to voice anxieties about randomness, they construct a lyric subject that worries about what might displace the expressive subject itself. Their discourse of anxiety suggests that a text open to the messiness of chance can prove more expressive than works that, like Stein's, seek autonomy and aesthetic enclosure. The *Stein* poems thus invite reconsideration of how lyric norms organize relations between expressive speech and historical information. Their anxious voice marks a generic difference from Stein, whose more composed response to contingency indicates a fuller departure from conventional strategies for managing meaning and value, and it denotes uncertainty about the generic and literary-historical position of the *Stein* poems.

Mac Low's lyric expression emerges from a rule-bound system of chance operations, rather than a conventional scene of authorial voicing, because such systems govern the historical scene to which his poems refer. What Derrida says of chance in linguistic systems applies more broadly to this scene: "Competition between randomness and code disrupts the very systematicity of the system while it also, however, regulates the restless, unstable interplay of the system."[91] In other words, the play of chance in various systems—technoscientific, geopolitical, literary-historical, compositional, hermeneutic—both structures and destabilizes the contexts in which the *Stein* poems make sense. Physicists systematize chance to make nuclear weapons, thereby producing both the reference book Mac Low uses and the anxieties about the chance of nuclear war that his poems seem to express. It appears less strange that lyric expression should emerge from the systematized play of chance if you consider, first, how deeply the chance of nuclear war informed psychic and social life during the Cold War and, second, how the instrumentalization of chance in various systems of social control helps to set the parameters of contemporary subjectivity, the locus from which a personal, expressive voice would issue.

The systematized play of chance thus reorients lyric expression and interpretation. Perhaps chance-operational writing adequately expresses the tenor of social situations that feel contingent, unstable. As Adorno warned earlier, though, chance operations might free the artwork from "the *pseudos* of a prestabilized harmony," but chance also leaves art arbitrarily formed, subject to the "blind lawfulness" of an accidentality that "can no longer be distinguished from total determination." He argues that "art does justice to the contingent" not by embracing randomness as such, but "by probing in the darkness of the trajectory of its own necessity," a probing that imagines its own necessity as it goes.[92] Allen Grossman similarly describes the lyric's claim to necessity: the lyric poem "is the one actual thing . . . that did happen in the situation at hand . . . a hostage in the one world where finally and unexchangeably the one thing that happens (the very thing) really comes to reside."[93] The expressive voice that calls out from the *Stein* poems can be understood as constructing just such a dark necessity, so to speak. Our resulting hermeneutic work rends textual details from the play of chance that animates the multiple systems from which the poems emerge, submitting these details to a necessity that makes interpretation possible. This *a posteriori* attribution of meaningful necessity to any detail is the critical gesture per se. Likewise in mathematics or physics, the next step in a random walk or the next swerve of a neutron seems arbitrary and weightless in advance but necessary and decisive when the calculation is complete, when the bomb goes supercritical. It is thus not enough to say that expression in the *Stein* poems emerges from the rule-bound play of chance because this reflects the systems of contingency that form the social world of the poems. In addition to this, the interplay of chance and necessity fundamentally structures the interpretation of expressive language.

Readers who believe chance operations tend to displace expressive lyricism often link this effect with broader claims that Mac Low's work intervenes upon aestheticopolitical norms. In this way, his work renegotiates the relations among personal expression, chance, and the historical contexts that influence readers and writers. Where his poems seem expressive,

they do not simply reassert old poetic modes with which Stein had dis-
pensed; they highlight the difficulty of literary evaluation and historical
interpretation in the context of procedural writing. Hence, when Jennifer
Scappettone discusses *Words nd Ends from Ez*, the rewriting of Pound's
Cantos, she notes that they "possess curious critical force," despite Mac
Low's procedural method.[94] She argues that Mac Low reconfigures Pound's
historical image, helping us reconsider the earlier poet's fascism, and she
claims that his chance operations in particular disrupt the causal logic
that normally guides historical sense-making. Discussing the same poems,
McCaffery finds in them a reading of Pound that not only "does add to our
information about the textual reality of *The Cantos*" but also "challenges
the basic premise of why and how we read." The *Stein* series precipitates
similarly "extreme cognitive disruptions that decommission conventional
methods of reading" while also illuminating Stein's work.[95] However, such
disruptions do not simply negate the familiar logic of poetic expression.
Scappettone notes that the "ego-trammeling, choral outbreak" of Mac
Low's rewriting "displaces Pound's imperious personhood" with a "lyricizing" voice all its own.[96] The ostensibly impersonal procedures of *Words nd
Ends from Ez* and the *Stein* series do not banish such "lyricizing" impulses,
then, so much as they guide readers to consider how personal expression
grounds political and historical readings.

Mac Low, for his part, acknowledges that chance operations tend to
displace personal expression, but for him, the technique provides a metaphysical rather than a political dispensation. He traces his interest in
chance operations to "certain Asian philosophical traditions," especially
"Buddhism, which regards the ego as an illusory formation."[97] His later
commentaries emphasize the importance of combining such impersonal
procedures with "personal" decisions, as he does in the *Stein* series. Discussing his difference from Cage, he writes, "There isn't such an overwhelming emphasis on nonegoity in mine."[98] In a 1995 interview with
Bernstein, he declines Bernstein's repeated invitations to make political
claims for his procedures: "I don't think the use of non-intentional methods has much relation to left-libertarian... emancipatory politics."[99]
Instead, Mac Low sees political value in other "works requiring performers
or readers to exercise personal choice," to make decisions that "act as analogies for libertarian communities in that performers make independent
choices throughout each performance."[100] By this view, chance operations
may tend to disrupt personal expression, but they do not promise some

anarchic destabilization of political systems. Mac Low would insist upon the functional *ambivalence* of a chance operation vis-à-vis any material history of its equipment or any possible expression of political will, a position that squares with his broader skepticism about historical-materialist politics.[101] For him, an aesthetics of chance is neither inherently radical and destabilizing nor inherently supportive of political hegemony and state power. Rather, it is as indifferent to political positions and historical contexts as to any expressive utterance that might issue despite, alongside, or indeed through a chance-operational process.

The interplay between procedural writing and personal expression informs the political meanings of the *Stein* series as well. Discussing his methods with Andrew Levy, the poet uses spiritual language to deploy an expressionist concept of creative invention:

> I come up with this kabbalistic idea of "saving the sparks." Saving the sparks is saving the creator's spirit—or whatever you want to call it. . . . Something like that is happening when one uses a book that was composed for some horrible reason. I've used the Rand Corporation's table *A Million Random Digits with 100,000 Normal Deviates.* . . . I have often felt that when I used the random-digit book, I was somehow saving sparks; you know, these people were good mathematicians, yet they were putting their spirit into these military projects.[102]

In a similar context, Scappettone interprets Mac Low's idea of "freeing the sparks" as a reference to Pound's poetic archaeology in *The Cantos*.[103] When Mac Low discusses Pound's politics with Bernstein, he argues that "the method used" to rewrite *The Cantos* "purges all that nonsense out of his invention," redeeming Pound's work from his fascism.[104] Given his wish to recover the "sparks" of personal invention, Mac Low seems to invest procedural writing with contradictory functions. On one hand, he believes impersonal procedures hold the redemptive power to cleanse a text of unhappy historical entanglements, such as Pound's fascist convictions or the "horrible reason" for creating the RAND book. On the other hand, the very creative "sparks" that he seeks to recover would seem to rely upon an expressionist notion of personal invention that the nonintentional procedure is expected to displace.[105]

The *Stein* series compresses this double motivation, for Mac Low directs his redemptive intentions not at the poet he rewrites (i.e., Stein) but at the spirit of the scientists who created the equipment for his procedure. In this respect, the *Stein* poems differ from *Words nd Ends from Ez,* which Mac Low frames as an effort to redeem Pound, and from other procedural series such as the Asymmetries and *Stanzas for Iris Lezak,* both dedicated to his first wife. Writing before the *Stein* series began, Louis Cabri argues that "it is often in the selection of source texts that the social conscience" of Mac Low's work "comes out," but the political meanings of the *Stein* poems emerge instead from the poet's equipment.[106] In 1998, when Mac Low began the *Stein* series, Gertrude Stein might have seemed little in need of redemption. Although the Stein works that he favors had mostly occasioned bewilderment and ridicule through the midcentury, the experimental writers and feminist critics of the 1970s and 1980s had done much to improve Stein's reputation. More recently, however, scholars have questioned Stein's radicalism, often on identitarian grounds. Historicist critics have noted that her lesbianism was relatively heteronormative, her Jewishness complicated by her collaboration with the Vichy regime, and her aesthetic innovations supported by inherited wealth and class privilege.[107] By noting Stein's recent political castigations, I do not mean to reassert the old cliché that her formal experiments are progressive despite her conservative politics, nor do I assume that her politics must register in her artworks. The poetics of chance indicates the impossibility of disarticulating the aesthetic from the political, but it also calls us to doubt facile harmonies between these spheres.

Other influential modernists, such as Pound and Eliot, come with comparable baggage, of course, but such accounts still raise questions about the enthusiasm with which experimental writers have cited Stein as an influence. Mac Low seems to have appreciated Stein since the late 1950s, and by 1978, he had joined other poets in writing praisefully about her.[108] His comments on the *Stein* series say nothing about redeeming her image, perhaps because he feels himself already too much *within* the development of her reputation. Yet Mac Low produces a reading of Stein, not just an imitation. As they read through Stein's texts, his chance operations provide a further formation of the historical chances that have brought Stein and Mac Low together within a trajectory of American experimentalism. By responding to such historical anxieties with a recuperative care for the production of the equipment through which he and Stein meet, Mac Low recognizes randomness not merely as destabilizing in relation to literary and political histories but as simultaneously formative of those histories.

By directing his reparative intentions at the equipment through which he rewrites Stein, Mac Low draws attention to the productive powers of randomness. Even more than the projects that seek to redeem a poetic personality, the *Stein* series enables us to "watch the spell being created" through the redemptive alchemy of his procedure, as Bernstein puts it, "without losing sight of the machinic principles through which it is engendered," since in this series, Mac Low ponders the origins of his equipment itself.[109] This concern suggests that the digits in the RAND book (and perhaps other tools of chance) do not provide a purely disruptive influence but evoke specific historical narratives. Through the RAND book, chance operations no longer appear fully opposed to rational thinking but instead make themselves available as supports for various kinds of knowledge production: as a means to solve physics problems, as a method of aesthetic creation, and as a way to render historical meanings beyond strict causalism. As I argued in the introduction to the present volume, the rhetoric of lyric exemption turns out to be remarkably fungible with the forms of knowledge it disavows. Poets may value randomness as an escape from strict rationalism and order, but they also anticipate its conversion into one or another deeper way of knowing.

Mac Low's chance operations, then, do not block critical interpretation any more than they block the emergence of an expressive voice. On the contrary, the *Stein* series takes chance as a basis for thinking through the relation between historical change and assignments of value. Already in Stein's discussion of the atomic bomb, contingency operated as a copula between historical thought and value. The same structure appears in critical discussions of Mac Low. For instance, Bernstein praises Mac Low's chance-based poetics, in which "the things of the world . . . are valued for themselves, without the intrusion of ego's desire for ordering."[110] In his attack on Mac Low's work, by contrast, Brett Bourbon coordinates the relation between contingency and value differently from how the *Stein* poems would have us do so. To open an essay that will "explain how we can read the concept of a poem by reading a poem," Bourbon cites a few lines from *Words nd Ends from Ez* and claims that "it is not poetry" because it conveys no idea of what a poem is.[111] Bourbon links this point with a broader concern about cultural critics who advance "the idea that poetry is nothing more than 'a name for a changeable set of desires and cultural ambitions'" (28). Judgments of value, he argues, outstrip the kinds of historical accident such cultural relativism supposedly privileges: "If it is true that every perspective is historically, that is, culturally and contingently, determined, then what choice do we have but to do what we do? . . . If we can make judgments, we are not merely historically determined" (31). Bourbon sustains the conventional opposition between, on the one hand, value judgment as a fiat of agentive rationalism and, on the other, the prison-house of relativism that leaves us overdetermined by our contingent historical positions. "If we must judge the contingency," he concludes, "either we judge our own contingent entanglement relative to our prejudices or relative to some further idea of what is good" (31). While the *Stein* poems do raise concerns about the "contingent entanglement" of value judgments and historical contexts, they also suggest a more productive response to this conundrum. Bourbon holds that a judgment bracketed within the contingent frame from which it emerges would be no judgment at all, so we can disregard such contingency's formative pressures on assignments of value—can and indeed must, if this absolutist theory of judgment is to remain intact. By contrast, the *Stein* series offers a reparative response to these contingent links between a poem's historical contexts and its value. It thereby suggests some affinities with other postwar and contemporary poetics of contingency, which Charles Altieri sees as helping "us develop somewhat different languages of value for the poetry's relation to its culture and to ours."[112] The *Stein* series acts as a foil to its source texts, redirecting readers to Stein's less anxious views, and as Mac Low repurposes the RAND book for poetic ends, he also explores the possibility of redeeming a book by detaching it from its original purposes. Where Bourbon seeks to bracket contingency as threatening to the discourse of value, Mac Low imagines recuperative strategies of reading and reevaluating, strategies that might prove less brittle in the face of such contingencies.

CONCLUDING HISTORIES

How, then, can a judgment of value be made about the *Stein* poems themselves, especially as readings of Stein's originals? One might value them simply by virtue of their relation to Stein, but that is precisely the kind of historical ligament whose contingencies the series explores. Hence, their value might lie in making explicit the vicissitudes of their own valuation and of the broader discourse of aesthetic value, especially in light of the questions about determinism, agency, and chance the poems also open.

For both Stein and Mac Low, chance and related ideas of disorder help to figure the uncertainties inherent in historical thinking, but their responses to such uncertainty differ significantly. Stein offers a composed response to even the most forbidding avatar of contingency, the atomic bomb. Instead of viewing uncertainty as troubling or disruptive, she argues that our mediatized cultures provide *too much* information. A more local, quotidian "common sense," she believes, will secure the discourse of value, of lived "interest." By contrast, Mac Low's rewritings express anxiety about "this unordered spectacle" they make of Stein's texts. Contrary to frequent claims that experimental poetry offers positive alternatives to an expressive norm seen as retrograde and lapsed, Mac Low explores what forms of expressivity still lurk within a writing practice that overridingly engages language as a mechanism for procedural manipulations. In his pursuit of a reading practice that will illuminate Stein's more placid posture and in his efforts to "save the sparks" of invention gone awry, he suggests that such latent expressionism may be central to his poetics and, perhaps, to other writing that seems at first to liquidate old literary modes altogether.

A Million Random Digits acts as a lynchpin connecting Stein and Mac Low. This strange poetic equipment simultaneously randomizes Mac Low's procedure and insinuates its own unhappy history. By providing both constitutive disorder and historical context, it enables readers to see chance operations not as simply disrupting interpretation but as opening new historical meanings. As a model for reading poetry through its compositional equipment, *A Million Random Digits* makes at least two additional contributions to new media poetics. First, the word "digits" in its title suggests that projects for reading "digital poetry" could further clarify the encounter with digits per se. There is little consensus about the ideality, determinacy, or exactitude of mathematical discourse, especially in discussions of randomness, infinity, and zero. The RAND book's profound numerical disorder might undermine efforts to position the text historically, since this positioning would be totally accidental with regard to the actual numbers in the book. A clearer sense of when mathematical absolutes do or do not apply may prove important to future work in "digital" poetics. Indeed, the privilege of information in much current scholarship on electronic poetry rests, more or less explicitly, on the presumed ideality of the digital.

Second, the RAND book's role as an apparatus of chance operations serves as a reminder that not all scenes of digital composition or machine reading are informatical. Most current theories of new media poetics understand computers as providing access to positively given information—

often to large amounts of raw, numerical data—that computers help us compile and manipulate.[113] By contrast, Mac Low's use of *A Million Random Digits* shows that digital equipment sometimes provides an absence or disordering of information. Poets and their readers may find that computers occasion uncertainty, frustration, improvisation, disorientation, and befuddlement just as often as they support a fully knowing instrumentalism, but we do not yet have comparable models for reading these less orderly, less informatical roles for our digital equipment. What might it mean to look at a computer or a book of digits not as a knowledge engine but as a mechanism of uncertainty, or as a machine that can think but not know? Such a reconfiguration of the digital humanities in relation to the economies of knowledge they produce may improve our understanding of how we write, read, and live with electronic machines.

The *Stein* series is most valuable, however, for showing how the poetics of chance provides a lyrical supplement to historical knowledge. The series indicates that indeterminacy in Stein's work and Mac Low's should not be seen as concretizing a definite lineage of experimental writers. Mac Low's chance operations instead occasion a whole range of historical affiliations and deferrals, none easy to privilege. Like many other poets, he embraces chance as a means of exempting the lyric from systems of rational causation—historical, physical, and others. But Mac Low's poems also challenge the assumption that chance operations make it possible to abjure personal expression or to efface the historical situation of the body writing. They thereby suggest that other chance-operational texts might take comparably subtle approaches to literary history and that the afterlives of lyric expression may remain discernible in Language poetry, conceptual writing, and other practices that ostensibly displace the personal voice. Through these readings that the *Stein* poems produce, the play of chance supplements historicist hermeneutics with a description of Stein's and Mac Low's poetics, a description of the generic, aesthetic, and formal conditions that enable their writing. Far from dehistoricizing their work, such a poetics offers new historical vantages. For Mac Low and Stein, chance not only structures a text's relation to the circumstances from which it emerges but also informs its future value. Both writers acknowledge that historical relations themselves can be undecidable or accidental without for that reason becoming less consequential. Through this engagement with the operations of chance, the *Stein* series pursues a reparative project that not only puts *A Million Random Digits* to less destructive use but also resituates Stein's work and its significance for today's experimental poets. The next chapter explores dyadic relationships similar to the one between Stein and Mac Low as it reads Frank O'Hara's apostrophic calls to anonymous friends and future readers, but it traces these calls through a more familiar electronic device: the telephone.

3 Anonymity

Frank O'Hara Makes Strangers with Friends

The most powerful lines of Frank O'Hara's poetry express a poignant sense of anonymity and social detachment. To say so, however, contradicts not only the majority of O'Hara criticism but also the facts of his life. He was remarkably social, often surrounded by the notable artists, writers, and intellectuals of New York City's postwar avant-garde. Larry Rivers famously said at the poet's funeral: "Frank O'Hara was my best friend. There are at least sixty people in New York who thought Frank O'Hara was their best friend."[1] Since then, O'Hara's reputation as a socialite has only increased. The anecdotes of his life come to seem as familiar as the poems, and scholars often read his poems in terms of his busy social life.[2] This interpretive habit obscures a counterdiscourse of anonymity and estrangement evident in many of his best poems, a more conflicted perspective on being social. Alongside the witty self-portraits and the vignettes of a busy social calendar, much of his work expresses an unexpected sense of alienation from his friends, his lovers, and even himself. As Keston Sutherland puts it, a "set of preoccupations with whom O'Hara knew . . . and how enviable his social life was, has played a good part in distracting attention and interest away from another fact about O'Hara's poems, namely, that they are full of anonymity." For Sutherland the "wounding" pathos of this anonymity marks the lyrical wish for an impossible social closeness, a utopian intimacy, but anonymity itself plays a more active role in constructing O'Hara's theory of the social than Sutherland recognizes.[3] The social condition of anonymity emerges in O'Hara's poetry as the lyrical other of a rationalized scene where everybody knows your name. Anonymity shapes O'Hara's reflections on a range of social equipment—what we now call "social media"—that interrupt and block the very exchanges they apparently sustain.

The opening chapters of the present book discussed writers who distinguish lyricism from rationalism in order to claim that poems ultimately support alternative ways of knowing. This rhetoric of lyric exemption produces new knowledge by first disavowing it. In this chapter, the exchanges between poetry and knowledge unfold differently. O'Hara does not celebrate anonymity but sees it as the necessary by-product of his poetry's interrogation of the technical ground of social life. He does not distance his poetry from critical thinking but takes it as a means to shed light upon the social equipment we normally ignore so that we can focus on the experience of being social. Anonymity emerges where O'Hara interrupts a social scene, rendering it inoperative to gain insight into its function and, ultimately, into the technical conditions of possibility for social experience. As a lyrical other of the critical insights O'Hara's poems provide, anonymity contrasts against the very discursive system from which it emerges. We might make similar points about earlier chapters, for there too, poets undertake *détournements* of electronics against the highly logical, rationalistic principles supposedly governing their function. O'Hara's poetry itself, however, remains more directly supportive of knowledge production while also providing lyrical accounts of technology's influence upon social life.

O'Hara most often achieves this effect through the technique of address, the call by which his poems reach out to someone. A great many of his poems call out to others, whether to friends by name, to celebrities, or to an anonymous "you," but the very figure that would seem to construct a poem's social relations enables O'Hara to dramatize the experience of anonymity and estrangement. He concludes "Morning," for instance, with this melancholy appeal: "if there is a / place further from me / I beg you do not go."[4] At first, this reads as a typical plea to a lover not to depart, but somehow the addressee is already as distant as the speaker can imagine. Even if the "you" is physically near, he remains in some way superlatively remote from the place of enunciation, perhaps too far to hear the call of the poem. Such address directs readers to think critically about the techniques of social life, and it describes the counterintuitive lonesomeness of O'Hara's social world.

O'Hara contrasts poetic address with another kind of call, arguably of equal importance to him: the telephone call. Throughout his writings, he presents the telephone as an ordinary piece of equipment for social life.

His poetic address points up the technical foundations of socializing, but the telephone appears as a ready-to-hand means to communicate. In everyday use, the telephone's technical nature recedes from view; we do not think critically about how it structures and mediates, facilitates and impedes, social experience. Recalling the Heideggerian distinction between technology and equipment established in the introduction of the present volume, the telephone is *equipment* because we ignore its technical workings in order to make practical use of it, and poetic address is *technological* because it interrupts social experience in order to lay bare the conditions of possibility for the social. Indeed, when we do hold up a device like the telephone for close scrutiny, instead of simply using it, we interrupt its normal function. To think deeply about the function of a tool, whether a hammer or a smartphone, makes it harder to use, so any functional equipment solicits a strategic disregard of its actual workings. Even "information technology" does not fully rationalize the scene of its use but rather, like other useful tools, continually eludes the very instrumentality it enables. Hence, the call of O'Hara's poems becomes visible as a technology only by refusing to sustain easy, unobtrusive social exchanges as a telephone does. When his writing stages the opposition between poetic and telephonic calls, it dramatizes an interplay between lived social experience and critical insight about sociality as such. The increasingly complex telephone exchanges of O'Hara's day promised to rationalize the structure of communications while also immersing users in a snarl of wrong numbers, missed calls, party lines, and long-distance intimacy. In turn, the calls of his poems give us insight into social technologies while also rendering social life nameless and detached. Through this latter exchange, one can better understand the conditions of possibility for social life but only by thinking critically about the technologies that bring us together, thereby interrupting their function. Anonymity is the symptom of such interruptions.

In O'Hara's work, the telephone stands in for an array of electronics that proliferated in the 1950s and 1960s, devices such as the portable tape recorder, television, and transistor radio. Telephony in the mid-twentieth century might seem distinctly "analog," since voices were still transmitted as electromagnetic waves, but O'Hara's telephone plugged him into a system then at the forefront of developing what would be known as "information technology." In 1947, the solid-state transistor was invented at Bell Laboratories, the research arm of AT&T, the company O'Hara and most other Americans paid for phone service. In 1960, the same conglomerate

began operating the first communications satellite, a huge Mylar balloon
that reflected telephone signals and other transmissions. In 1963, the company introduced push-button dialing, and in 1965 it activated an electronic
telephone switching system in New Jersey, the first large-scale electronic
computer used for networking. It would be easy enough to offer a narrative
leading from O'Hara's technological situation to our own, but today's technologies share more with his poetic techniques than with his ideas about
the telephone. While O'Hara frames telephones as everyday equipment,
many studying electronic sociality today describe the same effects associated with his poetic address: anonymity, isolation, and self-estrangement.[5]
O'Hara took great interest in the ordinary and mundane, especially in how
these fatigue critical thinking and make experiences less vivid.[6] Too often,
though, scholars respond to his jokey celebrations of the trivial by equating his poetic techniques with such everyday equipment as the telephone,
thereby flattening a distinction crucial to his poetics, eliding his interest in
anonymity and estrangement. While the telephone is for O'Hara so ordinary that we forget to think critically about it, his techniques of poetic
address bemoan the individual's isolation in the crowd and articulate an
impossible desire for some social experience less structured and mediated
by our technologies for calling others. If social life has only become more
conspicuously transformed by electronics since O'Hara's time, then it has
also become easier to see how his poetic address illuminates the technological conditions of social life. In turn, by paying attention to the equipment with which O'Hara contrasts his poetic address, it becomes possible
to displace some of the burden of critical thinking from his poems and
hear more clearly the social yearning that motivates them.

THE POET ON THE TELEPHONE

O'Hara makes the telephone's importance clear: he often poses with a
phone for photographs, and he mentions telephones in many of his poems
and other writings. His engagements with the telephone seem to invite
a common misreading that equates poems with telephonic discourse and
frames O'Hara as America's great poet of gossip and chitchat.[7] He may
certainly be that, but this account misunderstands his ideas about technology and elides the melancholic dimensions of his poetry. Even his photographs with a telephone suggest something other than the bright optimism
of social togetherness. With the handset at his ear, he makes a show of
directing his attention to the telephone rather than the camera, thereby

performing the literal sense of apostrophe, a turning-away from the audience; he poses for us but not toward us (Figure 7).

Perceiving that the telephone has something important to do with O'Hara's ideas about poetry's social meanings, many critics assume it positively analogizes a poem for him. The idea of a "verse telephone call" appeared as early as Marjorie Perloff's *Frank O'Hara: Poet among Painters* (1977) and remains commonplace.[8] Contrary to such claims, the telephone for O'Hara does not provide a simple metonym for the chattiness, triviality, ordinariness, or indeed the social connectivity of a poem, but quite the opposite: he defines the social potentialities and technical conditions of poetry by distinguishing these from his use of the telephone. The most famous and most often misread of his comments on the telephone appears in his short essay from 1959, "Personism: A Manifesto." Here he recounts the birth of Personism:

Figure 7. Frank O'Hara talks on the telephone in 1965. Photograph by Mario Schifano.

It was founded by me after lunch with LeRoi Jones on August 27, 1959, a day in which I was in love with someone (not Roi, by the way, a blond). I went back to work and wrote a poem for this person. While I was writing it I was realizing that if I wanted to I could use the telephone instead of writing the poem, and so Personism was born. It's a very exciting movement which will undoubtedly have lots of adherents. It puts the poem squarely between the poet and the person, Lucky Pierre style, and the poem is correspondingly gratified. (*Collected*, 499)

Critics often claim this essay cannot be taken seriously, so their equation of poems with telephone calls works as a kind of pop joke, vaguely ludicrous but still meaningful. Redell Olsen believes the poet's "tongue-in-cheek observation that he 'could use the telephone instead of writing the poem'. . . highlights O'Hara's interest in a poetics of ephemerality that might blur the distinctions between art and life."[9] Olsen is not alone: Perloff avers that "much of this 'manifesto' is, of course, tongue-in-cheek," citing the telephonic reference in particular.[10] In the same collection in which Olsen's essay appears, two others by Rod Mengham and David Herd refer to "Personism" as a "mock-manifesto."[11] Hazel Smith refers to the juxtaposition of poems with telephone calls as a "meaningful joke," claiming that it indicates a "correlation between poetry and live talk."[12] Perhaps these scholars believe a serious poetics should offer a more staid narrative of its own emergence. To the contrary, reading "Personism" as an earnest statement reveals a rich and challenging set of ideas about poetry's relation to the equipment of social life. The telephone, for O'Hara, does not analogize poetic address, figuring an easy transit between art and life; rather, its ordinariness, its status as social equipment, contrasts with the difficulties of social exchange his poems dramatize. Telephones link us with others so readily that their technical nature evades attention, but O'Hara's techniques of address call attention to themselves as technical, defamiliarizing the technologies that mediate social life.

The telephone thus provides a negative analogy for the call of O'Hara's poems, and their differences underscore the importance of anonymity and social distance in his work. Oren Izenberg perceives as much: "The realization that the poet could simply call his beloved on the telephone does not lead him *in fact* to call his beloved on the telephone."[13] If it had, we would have one poem fewer from O'Hara. Contrary to common opinion, O'Hara does not write poems to friends *as though* calling them on the telephone;

he writes poems to friends *instead of* calling them. This distinction marks the difference between a telephone call's easy social connections and the poetic call's difficulties with social contact. Personism, then, is as much a social theory of telephone calls as of poetry. It may "have lots of adherents," even though "nobody knows about" it, because unwitting Personists who prefer to gratify their lovers call them on the telephone and declare their love directly: "If they don't need poetry bully for them" (*Collected*, 498). As Izenberg suggests, if we "reject the account of O'Hara's poetry as fundamentally 'personal' in the sense of being a communicative act directed at a single loved person," then we come nearer to seeing why, for O'Hara, poetry is personistic, as it were, rather than personal.[14] The rhetoric of address that marks a poem as "to" a friend ultimately signals the difficulty of making contact with those whom we ought to know best, including perhaps ourselves. Indeed, O'Hara's poems of address often come between persons in a negative sense, figuring social life as interrupted by the very technologies that sustain it.

O'Hara sees benefits in this poetic interruption of social life, but he allocates them counterintuitively. What can it mean that rather than a person, "the poem is correspondingly gratified"? The distinction between poetic and telephonic calls duplicates the structure of triangulation found in apostrophe. Just as O'Hara writes a poem to someone instead of calling him on the telephone, so does apostrophe in general direct itself to an addressee apart from the actual audience, to someone or something that cannot hear the call. The felicitous apostrophe, in other words, does not gratify its particular addressee but satisfies the rhetorical norms of apostrophe and thereby, perhaps, gratifies the lyrical parameters of the poem itself. Through this challenging conceit, O'Hara suggests that thinking critically about the conditions of social life disrupts those very conditions. He writes of Personism that "one of its minimal aspects is to address itself to one person ... sustaining the poet's feelings towards the poem while preventing love from distracting him into feeling about the person" (*Collected*, 499). One's feelings about another person work as a lure, occasioning an address that itself becomes the object of the poet's feelings and critical attentions. O'Hara's address makes visible the social technologies, including telephones, whose influences we normally ignore so that we can focus on actually socializing. The use of these technologies, without close critical reflection on them, forms the social ground that his poetry illuminates. Interrupting social experience, the Personist might replace the question "How do I feel about Vincent?" with the question "How do I feel about being social?" Although a profound lonesomeness emerges from this metasocial attention, O'Hara's erotic figures for this poetics indicate that he also finds something deeply gratifying in the process.

In accordance with "Personism," the poems distinguish their poetic address from the facile contact a telephone provides. O'Hara begins "Nocturne," for example, with the complaint that "There's nothing worse / than feeling bad and not / being able to tell you" (*Collected*, 224). Already the rhetoric of address gets tangled. O'Hara cannot tell you how bad he feels, yet these lines do tell you that he feels even worse about his inability to tell you, an enunciative impasse paradoxically resolved by its own enunciation. By the end of the poem, however, O'Hara has distinguished poetic rhetoric from the practical factors that keep him from telling you how bad he feels. If he cannot tell you, this is not because triangulated address stymies his attempts but "Because you have / no telephone, and live so / far away" (*Collected*, 225). To tell someone how I feel, in other words, requires nothing more than a telephone call or an in-person meeting. If poetic address complicates such easy social exchanges, it does so on a register distinct from practical communication.

Even as poetic address critically interrupts the staging of social encounters in the poems, telephones continue to facilitate social exchanges outside the discursive space of the poem. Indeed, they enable O'Hara to gesture toward a social outside that the text itself cannot develop. For instance, "To Jane, Some Air" addresses Jane at the end when the poet asks, "do you miss me truly dear," but the poem's most dramatic address is an apostrophe near the middle: "Oh space! / you never conquer desire, do you?" (*Collected*, 192). Having learned in the opening line that "what we desire is space," we might understand space not simply as a distance keeping people apart but as the matrix across which our desire for others unfolds. The next strophe continues the address: "You turn us up and we talk to each other / and then we are truly happy as the telephone / rings and rings and buzzes and buzzes." Here, "turn up" may refer to an earlier line in which the poet and Jane want "To turn up the thermometer and sigh," or else to turning up the volume of a telephone, or to turning up in the sense of showing up for a meeting. If space remains the "you" here and causes Jane and the poet to turn up and talk, perhaps they turn up in person rather than by telephone. We cannot reliably discern whether the telephone rings and buzzes because the poet ignores it in favor of a more intimate meeting or, on the other hand, because it sustains a frenetic social life. The next lines ask, "So is that the abyss? I talk, you talk, / he talks, she talks, it talks." The telephone's noises denote the same abyss they traverse. The conjugation of "talk" seeks to accommodate the telephone's strange technicity within daily experience. Reading "you talk," we may forget the earlier address to space and gloss "you" as personal, as Jane. The continuation of personal pronouns, "he talks, she talks," encourages this reading, but this makes the final "it talks" all the more unsettling. What talks when it talks? Perhaps "it" is the telephone, or else space itself, the matrix across which our social calls unfold. When it talks, she (Jane) talks and he (Frank) talks as well. The telephonic exchange sets up a play of pronouns and calls within the poem while also constructing its exterior as a space of easy social exchange. Without the poem, the telephone simply makes us "happy" by putting us in touch, even if ignoring the "buzzes" in the background means we fail to think critically about the equipment that links us.

Both O'Hara's poems and his own recorded interactions with the telephone underscore its power to interrupt us, including its power to interrupt poetry. In "3 Poems about Kenneth Koch," the telephone again appears as an exterior to poetic discourse. The piece frequently addresses Koch, exclaiming, "O Kenneth Koch!" and asking, "Are you getting the beer, Kenneth?" (*Collected*, 151). These lines speak to him as much as about him. The closing, however, dispenses with address even as it uses the telephone to literalize the apostrophic gesture of turning away: "Gee, I'm really depressed. / My black back. And now the telephone. 'Hello. Kenneth?'" (*Collected*, 152). Responding to his "black" mood, the poet turns his "back" on the poem and perhaps takes consolation in a phone call from Kenneth. The final two words are punctuated oddly: in the days before caller ID, the telephonic "hello" sounded expectant, interrogative, as it does in a recording of O'Hara answering the phone, discussed below. And if the period after "hello" implies someone speaking on the other end, then the subsequent "Kenneth?" signals a lingering uncertainty about the kinds of contact actual phone calls sustain, leaving us unsure, as they often did, about whom we might find on the other end. The closing positions actual telephone calls as capable of interrupting poetic discourse with more concrete social exchanges.

This interruptive effect is not merely a product of the poem, for the telephone really did interrupt O'Hara at times. In an October 1965 interview with Edward Lucie-Smith, for instance, O'Hara speaks at length about art and life in New York City, but the dialogue ends abruptly:

L-S: Your phone!

O'H: I know it. Can you stop this record?[15]

As though a ringing telephone were not enough to end the interview through reference to its technological outside, O'Hara pauses before answering to ask that Lucie-Smith stop the tape recording the interview. For O'Hara, such technologies draw us toward the social exigencies of the actual world, in which all sorts of equipment structures and mediates our interactions with others. But in the practice of daily life, the effects of this equipment can be forgotten so that the social contact they sustain will feel more direct, less subject to the deferrals, interruptions, and uncertainties with which his techniques of poetic address linger. In other words, the interview with Lucie-Smith is just an interview, its reliance on electronic recording equipment largely pragmatic and forgettable, but its interruption by another technology of the voice, the telephone, punctuates their exchange in a lyrical way.

The telephone thus sharpens O'Hara's sense that technology necessarily structures and mediates social life. Ordinarily it functions as mere equipment for communication, and its influence evades our attention. But even a device as mundane as a telephone can defamiliarize our techniques of social contact. For instance, in a 1966 short film directed by Richard O. Moore, part of the "USA Poetry" series, O'Hara is writing a film script with Alfred Leslie when the phone rings. O'Hara not only continues to type as he answers but also calls attention to the series of technologies in which he is caught up, saying to the phone:

> This is a very peculiar situation because while I'm talking to you I am typing and also being filmed for educational TV. Can you imagine that? Yeah, Alfred Leslie is holding my hand while it's happening. It's known as performance.[16]

The caller, Jim Brodey, offers a phrase to include in his writing, to which Leslie says, "Write it in."[17] This might suggest that O'Hara does intermingle writing and telephone calls after all. Indeed, Michael Magee reads this scene as one "where O'Hara's analogy of writing to telephone conversation has become startlingly literal."[18] But such a claim ignores the profound strangeness of this scene: one does not normally write while on camera, nor take up spontaneous poetic collaboration over the phone. What was

Brodey calling about? Surely not to deliver his suggested phrase, "flashing bolt." Brodey's stomachache is the only other topic discussed before the film cuts away. O'Hara involves this telephone call within a complex of other technologies, such as his performance before a camera, his use of the typewriter, and even his reaching out for Leslie's hand, and he thereby highlights the multiple ways technologies shape social practices, including the practice of writing poems (Figure 8). Unlike its quotidian role in "Personism," the telephone as a tool for poetic composition seems unfamiliar, conspicuously technological.

This technological attention enables O'Hara to develop a practical poetics more attentive to concrete social situations. When he does embrace abstractions, he does so not in the name of philosophical insight but for the pleasures of escaping from reason. In "Personism," he suggests poets should "try to avoid being logical," but he also derides "negative capability" and other romantic abstractions: "Personism, a movement which I

Figure 8. Frank O'Hara turns away from the camera and from Alfred Leslie, off-camera left, to talk on the phone while he works at the typewriter. From "USA Poetry: Frank O'Hara," directed by Richard O. Moore.

recently founded and which nobody knows about, interests me a great deal, being so totally opposed to this kind of abstract removal that it is verging on a true abstraction for the first time, really, in the history of poetry" (*Collected*, 498). His opposition to abstraction seeks to achieve "true abstraction," because he takes the latter as a gateway to the pleasures of lyric exemption, of poetry's license to exceed the bounds of the rational. He claims that "the only good thing" about abstract ideas "is that when I get lofty enough I've stopped thinking and that's when refreshment arrives." O'Hara pairs this flight from logical thinking with a careful, sustained attention to the practical technologies of social life. Some view "Personism" as dismissing serious consideration of poetic technique, but what reads as a dismissal in fact is an insistence upon poetic technique as a matter of practical, rather than merely theoretical, importance. We should approach poetic technique not through metaphysical abstractions but in the same way we scrutinize the technologies that sustain social life:

> As for measure and other technical apparatus, that's just common sense: if you're going to buy a pair of pants you want them to be tight enough so everyone will want to go to bed with you. There's nothing metaphysical about it. (*Collected*, 498)

O'Hara does not purvey metaphysical notions of adequation between form and content, only practical ones. The pants of "technical apparatus" should not just fit snugly; they should bulge around their contents. There is "nothing metaphysical" about his analogy for technique, for both telephones and tight pants rely upon physics, rather than metaphysics, to put us in contact with others. As with the erotic image of the "gratified" poem as "Lucky Pierre," this metaphor of technique as a tight pair of pants emphasizes the dual pleasures of technicity and concealment, instead of privileging one. The pants of form, after all, are erotic both for what they show and for what they hide. While O'Hara values poetic address for drawing critical attention to the techniques of social life, he avoids a grand metaphysics of poetic form that might curtail the aesthetic, often erotic pleasures of poems.

In one of O'Hara's most telephonic poems, "Metaphysical Poem," the telephone provides such an easy means of communication that it conceals both the technical supports for social life and the extent to which we remain, despite appearances, distant from one another. Unfolding as the transcript of a telephone call, this poem excavates the "metaphysical" thinking that shapes social activity when we ignore its underlying technologies:

> When do you want to go
> I'm not sure I want to go there
> where do you want to go
> any place
> I think I'd fall apart any place else
> well I'll go if you really want to
> I don't particularly care
> but you'll fall apart any place else
> I can just go home
> I don't really mind going there
> but I don't want to force you to go there
> you won't be forcing me I'd just as soon
> I wouldn't be able to stay long anyway
> maybe we could go somewhere nearer
> I'm not wearing a jacket
> just like you weren't wearing a tie
> well I didn't say we had to go
> I don't care whether you're wearing one
> we don't really have to do anything
> well all right let's not
> okay I'll call you
> yes call me (*Collected,* 434–35)

Instead of simply mocking the metaphysical poets, this piece clarifies O'Hara's own idea of the metaphysical. Whereas many O'Hara poems mention the telephone in order to contrast it with poetic discourse, this one directly performs a telephonic dialogue. Despite the casual tone, the interlocutors have difficulty communicating. The first speaker asks "When do you want to go," and the second responds not with a temporal designation but with a spatial one, a misgiving about going "there." When the first speaker then shifts to spatial questions, asking where they might go, the second refuses to express further preferences. After some discussions of who should accommodate whom, the spatiotemporal shift reverses itself. In response to the first speaker's avowal that "I wouldn't be able to stay long anyway," the second turns this temporal matter into a spatial one: "maybe we could go somewhere nearer." The poem seems metaphysical, then, because the slide between spatial and temporal registers indicates that the speakers have not met at an agreed time and place but continue to negotiate times and places, the meta-physics of their potential contact. After a further digression about the first speaker's clothing, the interlocutors at last find a topic they can discuss cogently, but this topic is the decision not "to do anything" together after all. The event that occasions the discourse of "Metaphysical Poem" is the nonoccurrence of a social event. The speakers cannot discuss anything coherently except the decision not to meet. Instead of settling on a time and place to go out, they agree upon a "meta-physical" meeting, a future phone call that, if it turns out like this one, will again concern the metaphysics of where and when to meet. This future call may amount to another series of miscommunications, followed by another agreement to defer social contact. Hence, the contact telephones provide seems metaphysical in multiple senses, not only as an alternative to physical meetings but also as a means of contacting each other to arrange further contact, or not. Whereas O'Hara's poems of address contrast themselves with the actual social calls the poet often made, this poem's dialogue exhibits the "metaphysical" sociality the telephone enables, underscoring electronic equipment's tendency to render social exchange at one remove from itself.

Though "Metaphysical Poem" does not employ poetic address, it is structured around a call across a distance, and its speakers remain anonymous. This anonymity shapes the central metaphysic of O'Hara's theory of the social, under which he recognizes the mundane physical supports of social exchange but also traces the abstraction of particular individuals into nameless, distant strangers. The first-time reader of "Metaphysical Poem" may experience this abstraction in the delay between beginning to read and eventually recognizing that the poem is a dialogue. Especially without end-stopped lines, the two voices at first blend together, and they become distinct only as they approach their decision not to meet. In these ways, the poem prefigures the technosocial conditions of our time. The telephonic dialogue here is indeed colloquial, as casual as the language we now pass through email, voicemail, and text messages. Like telephones before them, these platforms make it much easier to communicate, and they share with midcentury telephony a utopian vision of universal reach, of rationalizing the total social body within an orderly electronic network. But the poem emphasizes that such networks enable us *not* to meet in person, that they develop new social forms by abstracting our names and locations. From the chaos of missed calls, party lines, wrong numbers, and prank calls to the activities online anonymity makes possible today, electronic networks bring namelessness and distance to the center of social experience. O'Hara's poetic calls scrutinize these social abstractions of self. They do not valorize the abandonment of self in the name of an oceanic social feeling, as Whitman's calls to the nameless "you" of his future readership arguably do, but confront the anonymity that complicates contact with those we ought to know best. For O'Hara, becoming social marks the necessary horizon of our inner sense of self, directing us to a threshold of anonymity across which we may reach others without knowing who they are for themselves or who we might be for them. Sutherland perceives that O'Hara's social poetics refers us not to names or persons but to an uncertain experience of the social that renders the lonely pathos of his apostrophic calls: "Not only are these people anonymous, but what matters most about them is their anonymity."[19] For O'Hara, to think about sociality means thinking about anonymity, where the absence or insufficiency of a name marks a mode of lyric exemption, a rhetoric of social uncertainty, that textures and sustains our calls to others. Detachments from specific persons mark not only O'Hara's poetic address but also his broader view on social experience.

Even when his poems describe the in-person meetings that telephones help us defer, O'Hara frames social experience in terms of anonymity and estrangement. Consider the closing of "Personal Poem," in which O'Hara bids farewell to LeRoi Jones after lunch:

> I wonder if one person out of the 8,000,000 is
> thinking of me as I shake hands with LeRoi
> and buy a strap for my wristwatch and go
> back to work happy at the thought possibly so (*Collected,* 336)

Some who discuss this poem focus on the "two charms" he mentions at the outset or on the watchband he buys at the end in order to emphasize his postmodern poetics of consumerism and triviality (*Collected,* 335).

Others are drawn to the middle section's discourse on literary taste: "we don't like Lionel Trilling / we decide, we like Don Allen we don't like / Henry James so much we like Herman Melville" (*Collected*, 336). However, the obvious question about this poem's closing remains unanswered. How can O'Hara doubt that a single person is thinking of him at the very moment when he shakes hands with his friend? Biographically minded readers will note that "Personal Poem" was written on the day O'Hara founded Personism, so this is likely the poem he wrote for Vincent Warren, the "blond" he mentions in the manifesto. Under this reading, the poem's "one person" should be glossed as "one person in particular," not as "any one person." This is persuasive so far as it goes, but in this case, O'Hara's self-involvement still contributes to his detachment from Jones. We can read "out of the 8,000,000" as emphasizing that O'Hara loves one particular person out of so many, but it can also support a more radical reading that subordinates biographical contexts to the rhetoric of the poem itself. By these lights, O'Hara wonders if any one person out of so many—including LeRoi Jones—is thinking of him at that moment. A powerful figure of the lonesome poet lost in the crowd thus emerges. Physical proximity and even tactile interaction fail to guarantee a minimum of social cohesion, for we can never know with certainty that our friends are thinking of us and not someone else, or perhaps themselves. Indeed, even as O'Hara shakes hands with his friend, he thinks not about Jones but about himself, specifically about whether anyone is thinking about him, O'Hara. In a bizarrely circular exchange between social distance and self-interest, O'Hara has reason to doubt that his friend is thinking of him precisely because, for his own part, O'Hara is thinking about himself and not about his friend as they say goodbye.

Within this tableau of O'Hara's distracted farewell to Jones, the poet obliquely reasserts a kind of social relation, but one that remains estranged. Whether "one person" means Vincent Warren or literally anyone, O'Hara thinks about himself in a social way, by wondering if another person is thinking of him. His self-concern is not inward and solipsistic, as normally befits a lyric poet, but routed through others, made social, and thereby made less reliable. His wondering recalls a classic form of social anxiety, the paranoid question one asks oneself: "What do they really think of me?" We can never get reliable answers to this nagging question, in part because asking it sustains self-concern more than social engagement. Even when I wonder if my lover is thinking of me, O'Hara suggests, I am wondering about myself and not my lover, thereby distancing myself from those I might engage more immediately.

Like a telephone call, a handshake is a common basis for social exchange. In the context of "Personal Poem," however, the handshake becomes conspicuous as a social technique, one as likely to mark detachment as engagement. Through the lonesome crowdedness of O'Hara's poetry, such quotidian practices as the telephone call and the handshake become visible as technologies, as supported by specific materials and practices whose sheer ordinariness normally obscures their influence upon social experience. By defamiliarizing social equipment, O'Hara's poems remind us that socializing relies upon these devices that render others distant and anonymous. As Oren Izenberg puts it, "not to notice that 'the poem' as defined by 'Personism' is in fact a rejection of communication (rather than the literary emulation of it) is also to miss the fact that O'Hara's announced topic in 'Personism' is not in fact particularity but abstraction."[20] Through the abstraction of those one hopes to know, social experience comes to feel obscure, its supporting technologies more visible. As so many O'Hara poems attest, one can never be sure that the person one hears on the telephone or the person who grasps one's hand is present for such exchanges in a meaningful sense. They may, like O'Hara, be so distracted by the very question of their own uncertain status for others that the social act becomes abstracted, empty.

As it contrasts with his poetic investigations of social life, O'Hara's approach to everyday communications equipment brings into focus the many anecdotes that mistakenly set up the telephone as a positive analogue for the poem in O'Hara's workshop. Joe Brainard offers one such story: "He got up as tho to answer the telephone or to get a drink but instead he went over to the typewriter, leaned over it a bit, and typed four or five minutes standing up."[21] This often-mentioned penchant for writing in the company of others or while on the phone has encouraged the view that a poem functioned for O'Hara as a casual instrument of social contact, an element embroiled in the social fabric of daily life, much like a telephone call. However, if we imagine someone working at a typewriter in the middle of a cocktail party, we in fact see a gesture of social isolation, a turning away from others in order to write or else, as in the Moore film, an effort to dramatize the strangeness of everyday social equipment. Taking a telephone call isolates one from one's guests in the same way as writing a poem, but only in the name of social contact across the distance marked by the prefix "tele." Of course, poems too are fundamentally social. Even the loneliest poem is made of language and does, in some sense, communicate. Some of O'Hara's poems likely began as casual notes to his friends, enabling him to construct a coterie audience, but today their calls to "you" address an audience of strangers whose names O'Hara cannot have known. Even in the time of his writing, as the next section will show, his poetic calls to "you" tended to complicate rather than affirm social experience. By contrasting such calls with the telephone, O'Hara makes clear that his experiments with lyric address trace how the technologies of social contact render others distant and anonymous.

CALLING NOBODY

When O'Hara's poems address a nameless "you," they do not evoke intimacy with this second person, nor do they seek earnestly to communicate with them, nor even do they bemoan a loss of contact. Rather, these poems of anonymous address show how the technical supports of our social calls tend, whenever we pay them attention, to interrupt and defer contact. His address directs readers to think critically about the technologies of social life, and it underscores the surprising lonesomeness of his social world. These poems express the pathos of an estrangement that lingers over the lost feeling of social contact their calls mark out. Especially when they address an anonymous "you," they unfold the indefinite spatial and temporal distances across which such a call seems never quite to reach its addressee, even as it does span those same distances to reach its reader.

His poems thus underscore a tension inherent in lyric address. On one hand, his address often resembles the practical, quotidian rhetoric by which we actually call out to others. On the other, it is a technique of high poetic ritual, articulating a "triangulated" structure through which the poem's audience intercepts a misdirected call that will never reach its ostensible addressee.[22] In most cases, O'Hara downplays the technique's ritualistic aspect, which some consider "embarrassing," in favor of a more casual style of address that he stages within everyday social scenes.[23] His poetry does not invoke the dead or call to inanimate objects as often as it marks a more familiar experience of social desire that lingers with its own nonfulfillment. If, as Ann Keniston notes, apostrophe in the postwar period enacts "a longing for a lyric mode that is . . . hopelessly out of date," then O'Hara's less embarrassingly ritualistic, more colloquial address transposes this longing into the social spaces his poems describe.[24] The failure and yet the persistence of this social yearning marks the lyrical pathos of O'Hara's work.

Those reading his address as a commitment to casual talk, however, overlook the sense of artifice his address conveys, an artifice that interrupts the same social connections it seeks to make. To some extent, all poetic address entails such interruption, since it stages the call to an improper addressee as an indirect way of reaching the reader. As Jonathan Culler notes, "to address someone directly—an individual or an audience—one would not write a poem," for doing so "introduces a certain indirection," an effect Culler takes "to be central to the lyric."[25] O'Hara's poems often read like casual notes to his friends, seeming to chronicle a social world, but his address also renders distant and anonymous those with whom he would seem most intimate. Like other scenes of lyric exemption in the present study, this rhetoric of anonymity supplements the kinds of knowledge it opposes. Namelessness and estrangement, for O'Hara, call attention to the social equipment that normally escapes notice, including the poem that hails an unknown future reader even as it seems to call out to a close friend. His address indicates that being social means ignoring the equipment that structures and enables it, equipment that makes social life seem artificial and empty when we think critically about it.

Even when the poems to "you" do not express a social yearning, the difficulty of reaching others still structures their discourse. "As Planned," written in 1960, addresses "you" in order to set at a distance both the poet's sense of self and the destination of his address:

> After the first glass of vodka
> you can accept just about anything
> of life even your own mysteriousness
> you think it is nice that a box
> of matches is purple and brown and is called
> La Petite and comes from Sweden
> for they are words that you know and that
> is all you know words not their feelings
> or what they mean and you write because
> you know them not because you understand them
> because you don't you are stupid and lazy
> and will never be great but you do
> what you know because what else is there? (*Collected*, 382)

The first three lines leave "you" undetermined; the second person's only distinct quality is "mysteriousness" to himself. The praise of vodka might apply to him alone or to a more universal "you," a synonym for "one." Doesn't everybody find that drinking eases the pain of self-estrangement? Don't *you*? Soon, though, notes on what "you think" and what "you know" suggest that "you" applies specifically to the poet. The absence of punctuation further motivates this transition. Line breaks provide a syntactic guide to the opening sentence, but the conflated syntax of later lines suggests an ongoing string of thoughts, perhaps an interior monologue. Of course, "you" is a strange way to say "I," but in addition to suggesting interior self-address, this technique underscores the poet's self-estrangement by figuring it as the distance between first and second persons. Once focalized on the speaker, the poem berates "you" as insensitive to peoples' feelings, ignorant of what his words mean, and mediocre of character: "you are stupid and lazy / and will never be great." If some trace remains of the opening lines' hint that "you" could refer to someone other than the poet or could refer to everyone, including his readers, then O'Hara now verges on insulting and thus further estranging his audience. By addressing a nameless someone, he plays on a slippage between registers of the first person: when he writes "you," he means to say "I" but seems also to indicate me.

As it turns out, the impression that "you" means someone other than the poet may be warranted, and in this case, the poem's address draws attention to the technological, material histories across which the call of the poem reaches us. Donald Allen's note to the poem reads: "Dated December 16, 1960, in MS x141. The poem is a reply to a poem by Bill Berkson of the same date. First published in *Paris Review* 49, summer 1970" (*Collected,* 549). If the poem seems to hail someone other than its poet, maybe it addresses Berkson, himself a poet and close friend of O'Hara. The references to "you" as a writer narrow the scope of address, since not all readers are writers, but Berkson's vocation fits this constraint. Even as an insider's exchange with another poet, the anonymity of "you" allows O'Hara to play at insulting his friend while outwardly expressing more socially gracious self-doubts. The poem clearly sustained a direct social exchange between O'Hara and Berkson, but even at the time of writing, through its slippery rhetoric of address, it resisted functioning simply as a versified note from one person to another. A similar effect emerges in one of the most famous American poems to an unnamed "you," William Carlos Williams's "This Is Just to Say." That poem's status as an occasional note of apology is undercut when the poet lingers on how "delicious" the stolen plums were.[26] Likewise, "As Planned" is more than a simple note to Berkson, because its "you" can refer not only to Berkson but also to the poet himself, to a universal "one," and to the reader. In the time of our reading, to emphasize the bland historical fact that O'Hara wrote this poem to Berkson is to overlook what its fascinating play of address tells us about O'Hara's theory of the social. While Allen's note names Berkson as the addressee, it also opens other uncertainties instead of resolving this poem's structure of address. Allen refers to the Berkson poem by its date of composition, usually a more esoteric fact than a poem's title, yet he leaves no ready means of finding the right Berkson poem. One strains to imagine how Allen might have known that "As Planned" responded to a Berkson poem and known that both were composed on December 16, 1960, yet somehow not known the title of the Berkson poem in question—or why, if he knew the title, he did not include it in his note.

When asked about these poems via email, Berkson could not recall which poem had prompted "As Planned." Like O'Hara's anonymous address itself, Allen's note about a poem by Berkson has become a fact that leads nowhere, its referent lost to history. Berkson admits both poems were "written where and when I am not yet sure," though he suspects that "it was at Larry Rivers' house in Southampton."[27] He identifies the phrase "as planned" as "an expression I would have known," comparing it to "FYI," another shared term, and he believes the phrase arose "apropos some social plan that didn't work out or anyway changed so that we were left musing over it." Even as Allen's note to "As Planned" suggests a poetic exchange between Berkson and O'Hara, tracing the poem's historical background leads to social plans broken and long-forgotten. As a reflection on "some social plan that didn't work out," the poem's historical mise-en-abyme situates it as the product both of an exchange with Berkson and of the failure of a social event; without either, the poem would not have come to be. Like other poems discussed below, the occasion for this poem's call to "you" is the nonoccurrence of a social occasion. The uncertain destination of its address raises questions about the relation between apostrophe and more practical uses of address, suggesting that this relation changes in response to the shifting histories through which poems reach their readers. The poem's very identity as a poem available to scholarly attention, not simply a note to Berkson, is predicated upon its historical uptake into contexts its address cannot intend, a slippage of the call to "you" through the displacement of the text, its continual positioning as a lyric poem rather than merely a message to a friend. This interchange between the historical origins and the generic destination of poetic address rests upon the construction of an actual audience of unknown strangers, an anonymous readership, as the necessary horizon of the poem's call to "you." The social forms imagined through such address coalesce around the tropes of anonymity that emerge from the multiple senses of audience the poem conveys.

In other words, O'Hara's poems often explore the relation between the decontextualizing energies of poetic address and, on the other hand, the specific histories through which the call reaches us. Another poem written to "you" in 1960, titled "Song," does not call us to consider the social context of its production but does proliferate figures of social distance:

> Did you see me walking by the Buick Repairs?
> I was thinking of you
> having a Coke in the heat it was your face
> I saw on the movie magazine, no it was Fabian's
> I was thinking of you
> and down at the railroad tracks where the station
> has mysteriously disappeared
> I was thinking of you
> as the bus pulled away in the twilight
> I was thinking of you
> and right now (*Collected*, 367)

One might expect a poem invoking "you" in more than half its lines to stage an exchange with this second person, but instead it describes a series of departures and missed connections. The piece begins with social uncertainty: if you did "see me walking by," then you said nothing, since I am unsure, or else you missed me entirely. Just as the refrain's debut in the second line syntactically straddles the first and third lines, so does its address signal at once the impulse to reach "you" and the failure of contact that occasions this address itself. The refrain also makes the missed connection at the garage more poignant. Not only can I walk by you unnoticed, but even if I am thinking of you at the time, we both might fail to note our proximity. The implied full stop in the middle of the third line, however, more strongly links it with the second: "I was thinking of you / having a Coke in the heat." O'Hara's phrasing tumbles similarly over the next line break, as "it was your face" solicits "I saw." Such enjambment develops a syntactic habit that reaches culmination in the dangling final line. As the poem unfolds, more images of failed contact occasion more apostrophic calls. I saw "your face" in a magazine, but in fact "it was Fabian's"—another subtle barb veiled by slippery address. I write to "you" and not Fabian, perhaps because his identity as someone I have seen but not contacted seems comparatively stable, whereas my call to "you" never quite reaches its mark. We soon relocate from an auto shop to a rail yard, but the station that would fix our location has "mysteriously disappeared." Among broken Buicks and disappeared train stations, what we have lost is not so much "you," whose absence the poet might lament and seek to recuperate, but the very coordinates of position and vehicles of travel by which "you" could be seen as present and then absent. By now, we might suspect that if "I was thinking of you / as the bus pulled away in the twilight," that does not mean either of us was on the bus. Far from invoking a simply departed figure, O'Hara's address lingers over the vacated possibility of social contact against which such absence and loss would measure itself. Instead of simply bemoaning disconnection, his calls to "you" ask how any social experience can be anything but missed, barred, incomplete, and thus explore a kind of "unrecapturable nostalgia for nostalgia" in the very rhetoric of address that seems to call out to others (*Collected*, 300).

More enjambed syntax in the poem's second half sets readers up for the resonant final line. The second instance of the refrain, in the fifth line, reads as an independent clause, so the following line, "and down at the railroad tracks," seems a fragment until the refrain in the eighth line completes the sentence. Yet the next line, "as the bus pulled away in the twilight," makes the prior refrain seem pulled away from the sentence it had completed. We could read either "and down at the railroad tracks / I was thinking of you" as a complete sentence or "I was thinking of you / as the bus pulled away" as a complete sentence, but not both at once. Instead of clumping syntactically, O'Hara's phrases disperse across their line breaks. They are, like O'Hara and his friends, less certain of themselves for the structures that separate them. The final refrain briefly clarifies by resolving the unsteady "as" clauses, but the final line conclusively disrupts the poem's syntax. Unlike the earlier line beginning "and," which also was a fragment, no refrain sweeps in to provide closure. The punctual "right now" draws out the temporal aspect of what had seemed a series of primarily spatial disconnections.

To what present can this "now" refer? If it indicates the time of writing, then the refrain's failure to appear in the present tense ("right now, I am thinking of you") suggests that the task of writing keeps the poet from thinking of you. Thinking of the poem as he writes it, O'Hara might recall that he *was* thinking of you and describe some missed connections, but he apparently cannot think of you "right now" without leaving off writing. The very act of writing the poem therefore counts among the factors of failed social contact it depicts; as he mentions in "Personism," he writes poems instead of calling friends. The poet's friends claim he often wrote poems in the midst of parties or while talking on the telephone, habits that critics view as imbuing his poetry with the social energies of his life. But the distractible end of "Song" indicates that such practices actually are anti-social, since they mean that the poet's friends do not have his full attention. As John Ashbery recalls, "it was very annoying, actually," to talk on the phone with O'Hara and begin to "hear the typewriter in the background."[28] In Ashbery's experience, O'Hara could not really be "thinking of you" while also writing a poem. If, on the other hand, "right now" refers to our time of reading, we encounter the same distances the poem describes. You could not have known the poet "was thinking of you" when he walked by you at the garage, nor can you possibly know he is thinking of you, the reader, "right now" as you read the poem. Whenever you read, it seems impossible to receive the message that O'Hara is thinking of you "right now," at the moment he thinks. O'Hara is dead, so he cannot think of you as you read his poem. But this impossibility recalls the deflated goal of contact set forth in the first line's question. Even as the poet lived, the absent "you" addressed there could no more appear and answer O'Hara's simple question than O'Hara himself, in the time of our reading, can think of you. Nonetheless, the poem issues both impossible calls together, and they imagine an unreachable "you" through the figures of social distance the poem sets forth.

In O'Hara's hands, these distancing effects might seem intrinsic to the rhetoric of address. Indeed, Ann Keniston views "postmodern address" as "downplaying the optimism . . . of traditional apostrophe . . . by foregrounding the absence of the addressee," precisely O'Hara's technique.[29] But in this way, O'Hara differs from others who make more optimistic uses of apostrophe. In "Howl," for instance, O'Hara's contemporary Allen Ginsberg ritualistically calls to Carl Solomon, "I'm with you in Rockland," evincing confidence in his poem's invocational powers.[30] O'Hara approaches the rhetoric of address more skeptically, and he thereby underscores difficulties that might embarrass apostrophe in a wide range of poems. He might point out, in other words, that Ginsberg is not with anybody in Rockland. He is buried in Newark, and anyway he met Solomon at a hospital called Columbia Presbyterian, not Rockland. Such considerations are admittedly marginal for interpreting "Howl," but they would be central to O'Hara's understanding of its address. O'Hara's calls to "you" reflect upon the social distances such address unfolds, drawing attention to the uncertain relations between the historical scenes of his writing and the futures of its reception. While his "you" seems often to indicate a friend or even himself, it also constructs an anonymous audience for his poetry, a nameless other person who might hear his call someday. These anonymizing, distancing effects of lyric address work in tandem with O'Hara's interest in the material techniques subtending his social life and poetic practice, techniques as simple as walking by someone unnoticed and as complex as making a telephone call.

MAKING STRANGERS WITH FRIENDS

As often as O'Hara calls to "you" in his poems, he also addresses specific friends by name. Titles like "A Letter to Bunny," "A Note to John Ashbery," "To Larry Rivers," and "For Grace, After a Party" frame these poems as casual notes to his friends. Many readers view such poems as transcripts of actual social events, and this interpretive habit underscores the strange power lyric address has to conflate the actually existing world with the worlds poetry constructs. Nobody suggests that the main thing to say about Robert Frost's "Mending Wall" is that it proves Frost knew how to stack stone walls, perhaps because that poem does not stage the actuality of its scenes through address. Although many of the poems O'Hara addresses to named friends undoubtedly emerged as actual vehicles of social exchange between the poet and those around him, surprisingly many of these reflect upon social disconnection. Even when a poem seems obviously to be a note O'Hara casually jotted off to someone, any value it has as a poem, rather than simply a partial transcript of a particular biography, stems from its ability to define itself as something other than a note, to abstract itself minimally from the specific social contexts of its production. As we have seen, O'Hara signals his recognition of this fact by insinuating the unknown future audience of his poems as a possible target of his address. In the poems addressed to named friends, he also scrutinizes the nature of a poem's link to its particular occasion, the event of its writing, in order to suggest that a necessary condition of poetic success is the poem's drawing a distance between itself and the local circumstances of its production, the same guise of lyric detachment historicists lambast as a misguided invention of presentist readers.

Much of the joy of reading O'Hara no doubt stems from the impression that his poems offer a thoughtful and clever perspective onto an unusually lively social world full of brilliant artists and intellectuals. As Allen notes, a great many of his poems can be fruitfully read as "a record of his life."[31] Whether one considers this an adequate strategy for interpretation may depend whether one aims, in the end, to understand the poet or the poems. Those more interested in the poet might look to a piece like "To Larry Rivers," which asks to be read as a note to the painter in response to his discussions with the poet:

> You are worried that you don't write?
> Don't be. It's the tribute of the air that
> your paintings don't just let go
> of you. And what poet ever sat down
> in front of a Titian, pulled out
> his versifying tablet and began
> to drone? Don't complain, my dear,
> You do what I can only name. (*Collected*, 128)[32]

Responding to Rivers's anxiety about not being a writer, O'Hara gives generously of himself to comfort his friend, just as Rivers recalls in his eulogy for the poet. O'Hara both praises his friend's paintings and derides poetry: "You do what I can only name." The poem clearly is a note to a friend. However, if it has enduring value as a poem for us today, it does so because it departs from its restricted status as an item of correspondence. Without entirely ignoring its origin, we might appreciate this poem as a performance of friendship in general, since it suggests that a good friend will offer negative remarks about his own vocation in order to comfort his companion. Or we might value this poem as contributing to O'Hara's theory of the relation between poetry and painting. In any case, if the poems that read like notes have value today as poems, this value emerges when they exceed the status of mere notes to friends. Otherwise, they would bear no greater interest than the rest of O'Hara's correspondence or, indeed, a record of his telephone calls. This poem seems to recognize as much, for it describes the value of Rivers's art in terms of its detachment from the painter. O'Hara expresses surprise that his "paintings don't just let go of you." By the same token, we might presume that a successful poem will "let go of" its poet, drawing a distance from the specific circumstances of its creator in order to stand on its own. To the extent that his poems addressed to specific friends are valuable as poems, they must exceed their status as items of a specific correspondence.

In other poems to named friends, O'Hara more pointedly stages a distance from those with whom he seems to communicate, and these gestures draw attention to the technique of poetic address as artifice. In "A Note to John Ashbery," for instance, O'Hara offers high and humorous praise for his friend's poetry, calling it "More beautiful than wild ducks" (*Collected*, 33). But the poem ends with a figure of detachment, as Ashbery's readers "crane over the wave" separating us from the poet and "gawk" at his greatness. A comparable sense of humor informs "To Richard Miller," which begins with a question, "Where is Mike Goldberg?" The poem speculates at length about Goldberg's whereabouts but says nothing else to or about Miller (*Collected*, 301).[33] This gambit might have seemed funny to Miller, who probably expected a poem with such a title to concern him, but if it remains funny today, it does so not as the record of a little joke among friends, told over fifty years ago, but as a commentary on the ease with which poetic address can go off course. Indeed, by reading it today, we extend this divergence from its ostensible recipient, a divergence O'Hara seems to anticipate, since he has nothing to say to the addressee except to ask him about another person. Elsewhere, O'Hara still more directly shows us that poetic address can dramatize social distance as powerfully as intimacy. He opens "For Grace, after a Party" with a bracing claim of disconnection from Grace Hartigan, whom the poem addresses: "You do not always know what I am feeling" (*Collected*, 214). Even those whom we address most intimately, the poem suggests, cannot always know how we feel. In his virtuosic reading of this poem, Sutherland matches O'Hara's sense of detachment with his own lyrical discourse of anonymity: "The social circle whose circumference is my utmost bliss is made of this necessary anonymity, when I am its poet."[34] By addressing named friends, O'Hara's poetry verges upon a certain namelessness, a subsumption into

social spaces in which names signify nothing or mark only the remains of
what once were thought to be the destinations of our social calls.

O'Hara's poems addressed to friends thus provide the tantalizing promise of a window onto a social world, but by highlighting poetic address
as technique, they also figure social detachment and anonymity. These
poems invite us to consider not only their relation to the names they pronounce but also the occasion of their composition, since they often seem
to record a particular social exchange. The poems disclose so many biographical details. Berkson, when asked about his poem that inspired "As
Planned," even suspects whose house the poets were visiting at the time.
It is possible to imagine developing a detailed social biography of O'Hara
by tracing the occasion of every poem addressed to a named acquaintance.
But, just as his poems of address interrupt the quotidian social exchanges
we expect them to sustain, so too does he complicate the notion of poetic
occasion. Indeed, the titles of two of his most famous books, *Meditations
in an Emergency* (1957) and *Lunch Poems* (1964), directly interrogate their
own occasionality, the first by asking what good poetic musings can do in
a true emergency and the second through a bathetic deflation of the occasion itself. Likewise, the temporal annotations in many of his "I do this,
I do that" poems, notations we might now call timestamps, complicate
the temporality of the occasion by insinuating a gap between the time of
writing and of the events described. Must occasional poems be written
in advance of the event they commemorate, for instance, or can they be
written retrospectively? Is there a difference between these and poems
framed as utterances first made spontaneously in the midst of the event
they mark? By scrutinizing the idea of poetic occasion, O'Hara opens his
poems of address onto the unknown futures of their reception, distancing
them from the specific social occasions that give rise to them. When his
call to a friend arrives before us as readers, the occasion of this call generalizes itself as an occasion to ask how such an indefinite call to the anonymous future can ever be occasioned.

One such unsettled occasion is Kenneth Koch's thirty-sixth birthday,
in 1961. For this occasion, O'Hara writes "On a Birthday of Kenneth's,"
whose first strophe reads like a semicoherent toast to the birthday boy:

Kenny!
Kennebunkport! I see you standing there
assuaging everything with your smile

at the end of the world you are scratching your head wondering what is
that funny French word Roussel was so fond of? oh "dénouement"!
and it is good (*Collected,* 396)

The poet goes on to recall times when "we admired you" and praise Koch
for having "shot out . . . way ahead of the Russians" as a writer, encomia
that further cast the poem as a gift or a speech made on Koch's birthday.
For multiple reasons, though, "On a Birthday of Kenneth's" should not be
read as a discourse produced or presented on Koch's birthday. It instead
situates itself as "on" Koch's birthday in the sense of "regarding" or "reflecting upon," thereby setting up a distance between the poet and the addressee
whose birthday comes to seem an improper occasion for the poem. Allen's
note to this piece tells us the manuscript is dated February 28, 1961, but
Kenneth Koch's birthday is February 27. Where the text comes closest
to directly verifying its time of composition, the evidence suggests it
was written not on a birthday of Kenneth's but on the day after. Moreover,
the poem's only mention of a specific occasion, in the penultimate line,
refers to the previous Saturday: "no wonder I felt so lonely on Saturday
when you didn't give your annual cocktail party! / I didn't know why." The
Saturday prior to the poem's composition was February 25, two days
before Koch's birthday, so even if it had been written or delivered as a toast
at the usual celebration, it would not have been on Kenneth's birthday.
Instead, the poem complicates the idea of a poetic occasion and its social
affordances.

The poem makes a "dénouement" from the occasional to the medita-
tive sense of "on," from the semblance of a social event to a "lonely" rumi-
nation on the failure of such occasions to materialize. O'Hara anticipates
this nonoccurrence in the penultimate stanza: "these days didn't add up to
a year / and you haven't had a birthday." If Koch had hosted a party two
days before his actual birthday, O'Hara might then have made such an
observation: Koch wasn't yet thirty-six, technically, and because the poet's
actual birthday fell on a Saturday the previous year, the time between
birthday parties might have been less than a year. But Koch had no such
party, leaving O'Hara to take the nonevent as an occasion for reflection
that casts O'Hara as both lonely and self-estranged. He complains not that
he misses Koch but that he "didn't know why" he "felt so lonely" on Saturday, when there was no party. His retrospection about a party that never
happened improves O'Hara's self-understanding but also reveals him to be

so estranged from himself that he cannot scrutinize his own feelings without reference to a social plan, a routing of his sense of self through the social that we have seen in other poems already. O'Hara writes about Koch's birthday one day late. Did he belatedly notice it in his calendar? Did he recently talk on the phone with Kenneth, who might have mentioned his birthday was yesterday? In any case he does not title the poem "On Understanding Why I Felt Lonely Last Saturday." Instead, the poem's occasional title marks a nonevent that enables O'Hara to address his feelings of isolation and self-estrangement.

O'Hara's techniques of address thus enable him to express uncertainty about himself and thereby to imagine the uncertain futures of his poems, including the anonymous others they will ultimately hail. The final section of an ambitious early poem addressed to V. R. "Bunny" Lang, "A Letter to Bunny," expresses anxiety about misdirected address: "When anyone reads this but you it begins / to be lost. My voice is sucked into a thousand / ears and I don't know whether I'm weakened" (*Collected*, 23). Notably, the poet does not write that he will be weakened *if* someone other than Bunny reads this but *when* someone does. The letter interpolates us indirectly as those possibly weakening O'Hara—possibly but not necessarily, because these lines again emphasize self-uncertainty. O'Hara does not know whether he will be weakened when his voice reaches the thousand nameless ears of his future audience. A few lines later, he comments on the letter's power to render him present and absent at once: "See? / I'm away now, but I'm here." The question addresses both Bunny, the letter's primary recipient, and any stranger who may read it. O'Hara is "away now" in multiple senses. The epistolary framework already implies his absence from the scene where Bunny might read this letter, as does our reading it instead of Bunny. In an equivalent countermotion, the presence of the text renders him "here" for Bunny and for us. Of course, Bunny and O'Hara are both "away now" in the mortal sense, but through such deictic formulations as "I'm here," O'Hara lingers with the impression of presence that the rhetoric of address can provide, though often accompanied by a sense of distance, including the distance of death. In this context, the final words of the poem, "Love, Frank," seem both formulaic and still quite resonant. Like the signature of any letter, this one traces the event of O'Hara's writing to a friend, but it also brackets the broader rhetoric of address by which the poem ultimately seems to anticipate its unknown future readers.

Without doubt, in this poem and many others, O'Hara's poetic address to friends often records very real exchanges between himself and those to whom he wrote poems, and tracing the biographical actuality of these calls can sustain a great deal of readerly pleasure, a sense of contact with O'Hara's personal mystique and immersion in his attractive social world. But many of the same poems complicate the rhetoric of address in order to set the poet at a distance from his friends, to question what social event might occasion his poetic calls, to figure his experience of self-estrangement, and ultimately to open the poems onto an indefinite future readership that necessarily remains nameless and unknown to the poet. The names that many of his poems call become emptied of significance: his friends appear not as specific personages but as vague figures whose most legible qualities are distance, inaccessibility, inscrutability. Their names themselves often deflate into the indefinite "you" by which the poet might also address an anonymous audience of future readers. By highlighting the poetic occasion, meanwhile, his poems mark lyric address as a powerful means of constructing the poem's futurity. The poet's call to a friend, when read decades after poet and friend have died, underscores the endurance of a text through the changing material and technological circumstances of its dissemination. Finally, O'Hara brings the resulting sense of social distance to bear on his own self-regard, articulating a sense of personhood as uncertain of itself as of the others who seem so difficult to reach.

PERHAPS I AM MYSELF AGAIN

As we have begun to see, O'Hara loses himself in the play of social distances his poems set forth. This sense of self-estrangement ranks among the most poignant effects of his techniques of address. It suggests that there may be no "you" to address at all, except as a further formation of an "I" that cannot reach itself. His techniques of address thereby develop a model of selfhood that can be productively counterposed with the more familiar image of O'Hara as a witty urban trickster positively engaged with his lively social world. To read these two models of the self against each other illuminates the dual queer pleasures of his work. To do so also clarifies how O'Hara reworks the lyric "I" as the expressive nexus of his poetry, underscoring its solitude by routing it through energetically social rhetoric.

The emergence of self-estrangement through lyric address is evident in "Morning," a poem this chapter's first section cited to show how the poet sets his addressee at an impossible distance. Before making that melancholic long-distance call, however, the poem suggests that any such address has more to do with the speaker's self-regard than with a real intention to communicate. The address to a beloved "you," under this reading, provides cover for an alienated discourse of the self. The poem begins by avowing that "I've got to tell you / how I love you always" (*Collected,* 30). These lines may seem concerned with the beloved, but in fact, they describe what the speaker feels compelled to do. O'Hara later conflates love with the most quotidian gestures: "I need you / and look out the window / at the noiseless snow" (*Collected,* 31). Needing you seems less a strong emotional impulse than a thing the speaker might do lightly and noiselessly and then finish doing, just before he looks out the window or parks the car. Perhaps his "Morning" blues will be gone by afternoon. Some lines do address the beloved without reference to the speaker—"what are you doing now / where did you eat your / lunch"—but these further isolate the speaker by sounding the silence as the questions go unanswered. Through the rhetoric of address that at first seems to direct the poem's attentions at a second person, "Morning" emerges as a poem of self-concern.

The poem tells less about the speaker's love than about his self-estrangement. O'Hara recognizes the solipsism of this posture: "it // is difficult to think / of you without me in / the sentence." This sentence, for a change, does not focus on the speaker except secondarily, as the one having difficulties. Instead, it describes the central problem of this poem's address and of O'Hara's poetic sociality more broadly, its frequent reduction to the orbit of an isolated first person whose sense of self cannot coordinate social relations. Aside from questions that go unanswered, only two instances of address in this poem are not bracketed by a reference to the speaker. O'Hara writes, "you depress / me when you are alone." But why should someone else being alone make one depressed? Perhaps "you" enjoy solitude. The lines would make better sense in the first person: I depress myself when I am alone. This version would fit tidily with the rest of the poem, and its attractiveness indicates a further consequence of the poet's self-concerned discourse of love and loneliness. Because poetic address remains for O'Hara inadequate to the longing for actual social intimacy it voices, "you" becomes another name for my own loneliness. By routing his self-concern through the second person, he figures an unhappy estrangement from himself. The other direct address to "you" appears in the poignant closing lines cited earlier: "if there is a / place further from

me / I beg you do not go" (*Collected,* 32). O'Hara does not lament a departure or even a social detachment his address hopes to bridge, but an isolation from another and from oneself that his address directly figures. If, as
Allen Grossman claims, "the person who speaks in lyric is always alone,"
then O'Hara's peculiar rhetoric of social isolation renders him a stranger
even to himself.[35] In one gesture, these lines distance the speaker from his
beloved and indicate that a further distancing would make the speaker not
only more lonesome for others but also less sure of himself.

O'Hara closes a later poem, "Mayakovsky," with a similar use of address
to figure self-estrangement, but this poem also makes it possible to bring
the poet's melancholic self-regard into conversation with his more affirmative views on actual social experience. The opening section expresses
homoerotic desire: "If he / will just come back once / and kiss me," then
the speaker might stop "standing in the bath tub / crying" and manage to
"put on my clothes / I guess, and walk the streets" (*Collected,* 201). In the
second section, though, the poet turns away from erotic yearnings and
retreats to poetry: "I love you, / but I'm turning to my verses / and my heart
is closing / like a fist." These lines make a clear distinction between poetry
and social life, the oft-elided distinction the present chapter reasserts by
reading the equipment of O'Hara's social life against his technologies of
poetic address. The remainder of the poem does not express the social
desires with which it opened but offers a discourse of self-estrangement,
one in which the speaker expresses not only a low opinion of himself but
also, perhaps worse, a self-obscurity or detachment from his own personhood: "Now I am quietly waiting for / the catastrophe of my personality /
to seem beautiful again, / and interesting, and modern" (*Collected,* 202).
The speaker sees his personality as a kind of disaster and lacks the power
to change either his character or his self-perception. In a striking departure
from first-person rhetoric, the poem closes by figuring self-estrangement
through a play of address: "It may be the coldest day of / the year, what
does he think of / that? I mean, what do I? And if I do, / perhaps I am
myself again." The crisis of identity reduplicates as it moves through multiple conditions of address. The question in its original form ("what does
he think of that?") could read as inward pondering or a query to an implied
"you." If the corrected form of the question ("what do I?") more strongly
indicates inward pondering, this clarification leaves in its wake a deeper,
more troubling confusion between "he" and "I." Even after this transition,
the addressee of these questions remains unclear: if "he" is the poem's "I,"

then who is asking whom about whose opinion of the weather? The final sentence takes this self-estrangement two steps further. The conditional, "if I do," throws the preceding questions into doubt by suggesting that the speaker, whether "he" or "I," might not think of the weather at all. Even if he does have some opinion, "perhaps" tempers the potential recovery of identity with a lingering uncertainty. One can never be sure, it suggests, whether one can be sure of oneself. Just as this poem's play between "I" and "he" signals the speaker's estrangement from himself and from any listener his questions might reach, so do many O'Hara poems modulate his sense of the self and the social through lyric address. His poetic calls figure the distances across which he encounters his friends and himself as strangers.

As a luminary of New York's postwar avant-garde, O'Hara might reasonably expect that even those at the periphery of his social ambit would recognize him, know his name. But he also seems to have understood that, by facilitating anonymous interactions, the social technologies of his era had begun to reconfigure the possible structures of selfhood. In his interview with Lucie-Smith, as the poet compares the New York art scene with those of Europe, he describes how the structures of his social world tend both to ratify a model of coherent, publicly recognizable selfhood and to deploy pluralities of anonymous social relations:

> If you have an American artist and you are to, oh, give a party for him. Then you would have a very wide range of people who are not, who may or may not know each other. In fact the person that you give the party for would be the only cohesive element which will link them all together.[36]

The scenario at first seems to affirm a model of stable, legible selfhood: the guest of honor at the party, around whom the social event convenes, knows everyone and is known by everyone. We might easily imagine O'Hara as this central figure, but his hypothetical party has a different point. Unlike similar events in European art circles, where coteries of mutual recognition predominate, the poet's social world contains so many strangers that he might not know everyone, even at a party he hosts. O'Hara positions himself among anonymous others, all but one of whom has only partial knowledge of the others in the group. This social network has a hub-and-spoke topology. As the only person at the party who knows everyone there, the guest of honor has the hub position: the party revolves around

him, as it were. Others at the party have their common acquaintance with
the guest of honor as a link between them, so their social interactions may
take place by way of him, through him.

The telephone exchanges of O'Hara's time employed the same hub-
and-spoke topology, with traffic routed through a central switchboard.
Like the structure of the party O'Hara imagines, these exchanges both
rationalize a social structure, enabling anyone to reach anyone else through
a central axis, and make social exchanges feel indirect, estranged. O'Hara
thematizes this structure in "A Sonnet for Jane Freilicher," in which he
makes a "long distance telephone" call to his friend (*Collected*, 61). Instead
of addressing Freilicher directly, the sonnet addresses "Operator Eighty-
one," who helps the speaker place the call. He implores the operator to
"loan / to pretty Jane" the "secret puissance" of telecommunication, by
which he might hear "that breath more dear than Fabergé." At the end
of the poem, he imagines Freilicher making a negative judgment about
him, though without addressing him directly: "'How closer than Frank to
the cosmic bone,'" he imagines her declaring, "'comes the bold painting of
Fernand Léger'!" Here again, O'Hara does not use poetry to pursue an
intimate exchange with his friend but instead to make visible the technologies through which he calls to her, including the triangulation of his call
through an anonymous third party, the hub called "Operator Eighty-one."
Both the guest of honor at O'Hara's hypothetical party and the telephone
exchange operator hold a stable, central position in a social topology that
renders other positions and identities in the social network marginal and
obscure. In both cases, the hub-and-spoke structure serves at once to
rationalize the social body, potentially linking anyone with anyone else,
and to estrange it, articulating a distance between people. The hub-and-
spoke as social topology finds its predecessor in the panopticon, which
Michel Foucault claims should not be seen as a specific prison architecture
but rather "must be understood as a generalizable model of functioning;
a way of defining power relations in terms of the everyday life of men."[37]
As with Jeremy Bentham's prison, the hub-and-spoke of the telephone
exchange or the artist's party not only enacts social discipline but also produces new forms of subjectivity.

The interplay between anonymity and identity continues to influence
electronic sociality today. On one hand, many online activities take place
only because participants remain anonymous. These include illegal file
sharing, hate speech and harassment, and many kinds of online sex play

and sex work. From the weightless (if naïve) feeling that no one is watching which web sites we visit to the more pointed technologies of anonymity such as Tor and virtual private networks, the proliferation of electronic networking seems to have made anonymous communication increasingly common. The case of anonymous online sexual activity proves especially helpful for clarifying how the poetics of anonymity contributes to O'Hara's politics of selfhood, since the rhetoric of anonymous sexuality draws so heavily upon the epistemology of the closet that gay men, including O'Hara, have so long negotiated. On the other hand, though online activity often seems anonymous, both criminal and state-sponsored surveillance technologies increasingly suggest that it is impossible to detach one's activities online from one's personal identity. Indeed, descriptions of the internet as a site of self-invention and identity-political affirmation take much of their power from the supposition that, if online activities are surveilled by corporations, governments, criminals, and even our friends, then we should carefully deliberate our online performances of selfhood. Many of the political contradictions of the present emerge from this tension between the apparent anonymity of communications networks and their power to reveal our identities. By considering these present tensions as we read the interplay of anonymity and identity in O'Hara's poetics, it becomes easier to adumbrate his politics of self-relation.

This chapter emphasizes O'Hara's estrangement from others and from himself in order to develop an alternative to the more familiar view of his persona as an affirmative performance, poised and self-assured. No doubt O'Hara took great pleasure in the habits of witty self-presentation discernible in his poems and other writings, in films and audio recordings of him, and in the anecdotal recollections of his many friends. Further, it matters a great deal for the sociopolitical meanings of his persona, and probably mattered a great deal for his own lived experience, that O'Hara was an out gay man in the United States at a time when countless other gay men endured more difficult circumstances, often negotiating more dangerous forms of estrangement and enforced anonymity than anything found in an O'Hara poem. That O'Hara could develop a public and affirmative sense of himself and his social world should be recognized and celebrated as the kind of victory from which redemptive gay histories can be made. By this light, his affirmative strategies of selfhood might amount to a kind of queer flânerie: the poet is both distinctive and able to disappear into the crowd, discerningly tasteful but not so different from the rest of us, unattached enough to float from one urban tableau to another but engaged enough to take aesthetic, erotic, consumerist, and epicurean delights from each.[38] Among American poets, the ur-figure of such queer flânerie is Walt Whitman, who takes similar poetic pleasures in his movements through the city and his self-presentations. Through his own frequent address to a sometimes particular, sometimes universal "you," Whitman imagines his union not only with his countless anonymous neighbors in New York City but also with another crowd, the multitude of unknown future readers his "you" addresses today. Though O'Hara's address has less felicitous effects, his more affirmative strategies of self-presentation and his bemused observations of strangers in the city closely recall Whitman. Indeed, the keenest delights of reading O'Hara may stem from these more playful, Whitmanian poses he strikes. It would be absurd to claim that O'Hara's poems are pure artifacts of isolation and self-estrangement. They are objects of language, fundamentally communicative. Many of them did circulate as occasional notes or inside jokes among the coterie of friends and collaborators who were his first audience. Even today, some of his most enjoyable poems do not interrupt social experience but chronicle it, communicating to his readers a playful, affirmative sense of the poet's selfhood and his social experiences.

These brighter valences of O'Hara's self-conception seem essential to the poetics of queer personhood in the postwar United States, but the alternative discourse of the self as anonymous and estranged has equally powerful implications for contemporary technopolitics, as I have suggested, and for the politics of queer life. The political significance of anonymity has been addressed nowhere more effectively than in the queer theories of anonymous sex, a practice that offers not only singular erotic pleasures but also means of resisting normative demands for a stable social order. In other words, fucking a stranger can be seen as the opposite of showing a cop your ID. As Michael Warner puts it, those having anonymous sex "may be seeking . . . a world less defined by identity and community than by the negation of identity through anonymous contact."[39] Even more than having sex with a stranger, there is something sexy about being that stranger, about losing oneself in the veil of anonymous otherness. Especially for closeted sexual minorities unable to identify publicly with their preferred sexual expressions, being "no one" makes it possible to have the kind of sex one wants. Such self-loss may not open onto a transcendent, oceanic feeling of cosmic oneness, a social union across time and space so prominent in Whitman's work. It may instead stand as a rejection of the solace of self-recognition, a refusal to be rendered into a social order by giving one's name, and an effort instead to lose one's identity and the coercions of social status and futurity that attach to it. Of course, this is quite unfamiliar terrain for the biographical Frank O'Hara, who avoided anonymous sex. There is evidence that O'Hara sometimes encountered casual homophobia, but he moved in circles that enabled him to become a minor literary celebrity without concealing his sexuality—not a common experience for gay men in the postwar years. Biographically, then, O'Hara is not a suitable representative for the queer politics of disidentification and self-estrangement sketched above. Nonetheless, if this chapter has offered good reasons to draw distinctions between O'Hara's poetry and his lived experience, then perhaps the figures of anonymity and self-obscurity in his verse can be said to represent those darker pleasures less evident in his life.

We can therefore discern two strategies of self-regard in O'Hara's poetry, each with its distinct sociopolitical logics. On one hand, many poems paint the familiar picture of O'Hara as a popular, widely celebrated fixture of New York's postwar avant-garde. This image offers an affirmative figure of gay life and equips readers with a vocabulary to appreciate the poems as witty self-performances in which the poet engages thoughtfully, playfully, often joyfully with his social world. The predominance of this perspective stems in part from the fact that it squares nicely with O'Hara's actual life. On the other hand, some of the most moving lines of his poetry express a poignant sense of anonymity, isolation, and self-estrangement, often through the rhetoric of address. These moments cast the poet as melancholic and lonely, equally unsure of himself and of the others who seem so impossibly distant, so nameless. If this perspective yields a correlative politics, it is a politics of negation that refuses the redemptive narratives of social cohesion, self-discovery, and expression. Although the poet did not espouse such commitments in his actual social life, his poetry evinces a keen awareness of the poetic and erotic pleasures that the technologies of anonymity afford. The interplay between self-disclosure and self-obscurity recalls O'Hara's statement in "Personism" about poetic form: "If you're going to buy a pair of pants you want them to be tight enough so everyone will want to go to bed with you" (Collected, 498). Like O'Hara's poetic persona itself, the analogy is at once jokey and quite serious. Sexy clothing is sexy for both what it shows and what it hides. By the same token, there is social gratification in the experience of mutual intimacy that O'Hara's coterie no doubt enjoyed, but his poems also imagine a lyrical intensity in the experiences of anonymity and self-estrangement, even if these do not seem operative in O'Hara's biographical experience. Although the pleasures of O'Hara's more playful, affirmative performances of self cannot be ignored, an equally important, complementary mode of self-regard emerges from those moodier lines where O'Hara seems estranged not only from his friends but also from himself. Through the rhetoric of address, O'Hara links this pathos of self-estrangement with a quintessentially lyric tradition of namelessness. As Allen Grossman points out, "the speaker in lyric by contrast to the speakers in drama . . . and the speakers in epic narrative . . . has in effect a sponsor (the author) but no name, is prior to or posterior to name, is an orphan voice."[40] An adequate picture of O'Hara's poetic sense of the self and the social, not to mention his sense of relation to the lyric genre, must attend the interplay between these two pleasures the poems offer.

HISTORIES AND FUTURES

The rhetoric of lyric exemption in Frank O'Hara's work follows a structure different from that seen in earlier chapters. In those cases, poetry's value gets defined in terms of its difference from rationalism and knowledge. Poets and their readers then claim that modes of thought such as affect and chance, having been situated beyond the closure of rationalism, can in fact support alternative ways of knowing. Hence, for example, critics take affective response as a privileged way of reading because it is not subject to evidence and logic, but they end up claiming it sustains new modes of certainty. Lyric exemption functions differently in O'Hara's work, however, because he takes poems not as a means to escape rationalism but as technologies of critical thinking. His techniques of address distinguish themselves from the everyday communications equipment we use to socialize. The privileged example of such equipment is the telephone, which puts us in touch with others so easily that—or rather, so easily *because*—we do not think critically about using it. In O'Hara's poetry, by contrast, social calls function technologically, drawing attention to the commonly ignored ways our social techniques structure and mediate our relations with others and with ourselves. His poems thus sustain rational scrutiny of communications technologies, but in the process they also produce anonymity, the sense that those whom we would hope to contact have become nameless strangers, distant and hard to reach. The fact that his poems generally avoid responding to this experience of anonymity by attempting to confabulate some utopian promise of social intimacy stems, perhaps, from his recognizing anonymity as the price to be paid for critical insight into social technologies. Although poems work for O'Hara as instruments of critical insight, his poetics of address sets forth its own rhetorical exchanges between knowledge and its lyrical others. Because paying attention to social technologies interrupts actual social experience, his poems suggest that our social lives operate through a field of equipment we necessarily do not think critically about and that to scrutinize this equipment closely, to regard it technologically, means interrupting its function. Correlatively, by highlighting the function of communications technologies, O'Hara's poetry produces lyric exemption at another register, that of identity, as those one hoped to contact become nameless and distant. In O'Hara's case, poetry thus remains the hero of an ongoing exchange between knowledge and its others, but that exchange operates differently here than in the cases discussed in previous chapters.

Telephones act as a foil for O'Hara's poetic techniques. Other notable writers of his time made efforts to involve audio technology directly in their compositional processes. William S. Burroughs and Brion Gysin, for instance, produced audio tape cut-ups throughout the 1960s, and 1968 saw the publication of Andy Warhol's *a, A Novel,* a transcription of taped conversations between Warhol and Ondine, one of his Factory disciples. O'Hara's poetry would be appropriated for such experiments as early as 1968, when the artist John Giorno created *Dial-A-Poem.* This work made it possible, by dialing a New York City phone number, to hear tapes of contemporary poets reading their work, poets including Ginsberg, Ashbery, and O'Hara. During his life, however, O'Hara not only eschewed such experiments but made much of the distinction between his poetic techniques and such ordinary social equipment as telephones and tape recorders. The calls of his poetry bring critical attention to the technologies mediating social experience precisely because his poems complicate and interrupt the social practices that everyday electronics sustain without calling attention to themselves. In this way too, O'Hara differs from the poets and critics discussed in earlier chapters. Both Dickinson's respondents and Jackson Mac Low celebrate poetry's exemption from rational constraints by bringing it into contact with electronics. O'Hara, by contrast, articulates the value of poetry by distinguishing it from the everyday function of electronic equipment. He separates poetry from electronics not because the latter rationalize and organize discourse, as so many assume, but quite the opposite, because poetry offers a critical perspective upon the social equipment that normally eludes rational scrutiny.

In this respect, O'Hara offers productive lessons not only on the poetics of address but also on our strategies for thinking about electronic equipment today. Recently, humanities scholars have taken a great deal of interest in poets who actively experiment with new media and in the new research methods electronics enable. Much less attention has been paid, however, to the increasingly ubiquitous and ordinary equipment that reshapes our lives in language more subtly. This book itself, for example, was composed using MS Word, and many of the texts it cites were accessed online, saving me trips to the library. With notable exceptions, scholars have shown more interest in the promises of digital scholarly forms that may displace the academic monograph (forms no doubt worth developing and discussing) than in addressing how everyday word processors and web browsers have transformed the work of literary critics, poets, and everyone else.[41] By calling attention to the necessarily inattentive ways we think about the equipment of everyday life, O'Hara's poetry suggests that we might benefit from scrutinizing more closely the technology we use so casually. And by dramatizing how such critical thinking interrupts the function of our social equipment, his poetry also asks whether rational attentions of this kind have a necessary horizon, since we render social equipment inoperative when we take it up for close scrutiny.

O'Hara's techniques of address also enable him to respond to long-standing lyric norms often articulated in terms of solitude and sociability. From the sixteenth-century entrance of "apostrophe" into English to current archival research that blurs the line between poetry and correspondence, the possible types of lyric address have continually shifted, as have their social denotations. Despite the changing social and technological foundations of lyric address, some common effects have emerged. For instance, calling out to someone or something in a poem produces a structure of triangulation by which the poetic discourse appears misdirected or intercepted, reaching someone other than the ostensible addressee. This effect appears in a striking range of poems: we are not Freilicher, whom O'Hara addresses, any more than we are Marvell's coy mistress or Shelley's west wind. The triangulation makes each of these poems seem "overheard," not directed at the actual listener. John Stuart Mill proposes this sense of overhearing as a quintessentially poetic effect: "Eloquence is *heard*; poetry is *over*heard."[42] In the same essay, Mill famously claims that "poetry is feeling confessing itself to itself in moments of solitude," so he associates poetic overhearing not with apostrophe but with another rhetorical technique: "All poetry is of the nature of soliloquy." Nonetheless, because triangulated address gives the effect of overhearing just as Mill's soliloquy does, it becomes evident that both apostrophe and soliloquy, two techniques so influentially proposed as markers of lyric intensity, in fact share a common effect.[43] In a superficial way, apostrophe seems motivated by an impulse to communicate, to call out to another, while soliloquy connotes an inward turning. As O'Hara's lyric address amply demonstrates, however, an ostensibly social gesture in language can powerfully figure isolation and self-concern.

Some scholars misguidedly restrict apostrophe to addressing dead or nonhuman things. This narrow view stems from historicist perspectives that privilege the actual life of the writer, including the social relations we might hope to trace in his address to real acquaintances. Here again, though, we miss a great deal if we read O'Hara's poems addressed to friends primarily as documents of actual communications. His poetic address often complicates and interrupts precisely the communication it seems at first to pursue. Indeed, the greatest lyric intensities in his poems hinge specifically upon the detachment of his social calls from their particular historical originations, upon the rhetorical anonymity of even their most intimate addressees, and upon their general exemption from a rational, legible social structure. O'Hara's work does solicit historical attention at another register, however, because his poetic address could not function as it does without playing counterpoint to the electronics he cites as a quotidian outside of poetic discourse. His address, then, is fundamentally shaped by the moment of technological history from which he writes, and the continued unfolding of that history has no doubt continued to reshape the available modes of poetic address and their social meanings. The rhetoric of address, whether apostrophic, epistolary, or otherwise, remains a significant resource for many poets today, and the gains they take from such techniques often depend upon their ability to explore how new technologies enable new kinds of social calls.[44] Future studies of poetic address will benefit from continued attention to how these ancient techniques respond to the changing social equipment through which we call one another.

O'Hara's poems offer equally instructive perspectives on anonymity, its literary and political functions.[45] When his lyric address draws attention to the social technologies we normally ignore, social life comes to seem grounded in anonymity, never fully reducible to a rational order of named persons. His poems produce anonymity by staging a transition from one mode of uncertainty to another, from the everyday experience of socializing with friends whom we know, while ignoring the equipment that puts us in touch, to the poetic experience of scrutinizing the technicity of social life, which renders anonymous those whom we thought we knew. This drama of social detachment shares much with the romantic poets' anxieties about the self's relation to the social, but anonymity does not always work this way. In other contexts, it marks the site of historical loss, acting as a pseudonym when the author of a text is unknown or forgotten. Anonymity can also designate a strategic concealment of identity, a way to avoid the consequences of becoming personally associated with a given text or action.[46] Today, when an influential activist group names itself "Anonymous," strategies of intentional anonymity seem closely bound up with contemporary political and technological circumstances. Some argue that the social condition of anonymity, strategic or not, is a sign of modern urbanism or global capitalism.[47] In one of her last writings, however, Virginia Woolf argues that anonymity marked the archaic poet's happy communalism, a sense of social union ultimately fragmented by that quintessentially modern technology, the printing press, and by the modern ascendancy of the named author.[48] Whether anonymity seems a welcome affordance or a sign of social decline, it remains politically salient in one way or another, and its possible configurations develop in response to the changing technological environments through which a name might go unknown. This chapter has confined itself to describing one such environment in which anonymity provides a poet and his readers with a rhetoric of lyric exemption, but this account also joins a growing effort to develop broader theories of anonymity's technological, political, and literary meanings.

The specifically political stakes of anonymity have become especially evident in recent years. Cultural authorities in the West have become increasingly tolerant about public expressions of nonnormative sexuality and gender identity, perhaps meaning fewer people feel they must love anonymously or not at all. But, as the discourse of gay liberation has amply confirmed, the epistemology of the closet sets in motion an interplay of concealment and disclosure that remains central to our sociosexual lives. Lots of erotic desires remain taboo (sadomasochism, for example, and bestiality), and those so compelled may therefore pursue these anonymously, whether online or off. The internet also makes it possible to remain effectively anonymous while expressing political dissent on a global scale, even where such dissent is illegal. Ongoing debates about privacy and intellectual property, meanwhile, would lack much of their current energy if not for the prevalence of illegal file sharing and other anonymous online practices. Anonymity has even become a factor in discussions of monetary policy, as governments develop responses to Bitcoin and other cryptocurrencies that support anonymous transactions. Even as successes for gay rights in the West have made the negative politics of anonymity and disidentification less attractive to queer communities, the political power of namelessness remains quite evident in other arenas. The political vocabularies of anonymity, concealment, and dissociation developed in queer theory offer an important and underutilized resource for understanding how anonymity works in other contexts. It is possible to read Frank O'Hara's poetics of anonymity in relation to these broader questions about the social effects of namelessness, especially because his techniques of address respond to the electronics that directly preceded today's communications networks. Future study of O'Hara might continue to explore what his poetry can teach us about the electronics of his time and our own.

4 Improvisation

Amiri Baraka, Allen Ginsberg, and Spontaneous Poetics

The present study began with a chapter about internet archives of manuscript images. It closes with a chapter that explores comparable archives of recorded poetry readings. These two types of online archive, visual and audio, began to emerge around the same time, roughly between the mid-1990s and the mid-2000s. The best known audio archives of poetry are UbuWeb and PennSound, which appeared in 1996 and 2005, respectively. The Internet Archive also hosts a great deal of recorded poetry. One can also find recorded poetry on YouTube, Spotify, iTunes, and other commercial platforms. Just as the online Dickinson archives make it easier to see the poet's manuscripts without traveling to libraries in Massachusetts, so do these online sources enable us to hear a poet read without attending a live event. Before such resources existed, it was possible to buy recordings of poetry on CD, cassette tape, vinyl, and other formats. The cost of manufacturing and distribution, however, meant that only select performances by prominent poets could easily be heard this way. Until recently, most audio recordings of poetry remained in academic libraries and the private collections of enthusiasts. Online archives like PennSound have made it much easier to hear a poet's voice.

This term, "voice," carries a dizzying series of connotations that complicate the effort to describe how poetry gets heard.[1] In her poems, a poet can construct a persona who speaks in a certain voice, a tone or dialect. Voice can also refer to prosodic habits or to the poet's physiology, how she sounds when she speaks. More broadly, voice signifies a tonal gestalt that makes a poet's work "sound" a certain way; when we say a poet has found her voice, this is what we mean. Voice in this sense is not exclusively aural.

It also ensures the poem's expressive authenticity and lyrical intensity, rather like Dickinson's handwriting, which supports feelings of intimacy with the poet. As Lesley Wheeler notes, these broader "tropes of voice . . . often mark challenges to how lyric poetry is defined, and, paradoxically, references to voice in critical writing can signify a resistance to the rational process of criticism itself."[2] Even as such vocal tropes mark the lyric's exemption from "rational process," they also obscure the technological systems that make poetry audible. Electronics preserve the poet's voice, but they also distance the listener from live performance, where direct hearing lends the voice its aura. The interplay between the poet's voice and its technological reproduction changes how we experience and interpret poetry.

These tensions between the poet's voice and its equipment have especially significant consequences for poetic improvisation. When it becomes easier to hear poets read, we notice improvisation more often, for spontaneity is more discernible in the recording of a live performance than on a printed page. In turn, to listen for improvisation in online archives calls into question what constitutes a poetic text and how to interpret it.

Literary critics commonly assume the texts we study remain reasonably stable over days and years, an assumption that seems incompatible with the idea of improvisation. Of course, this presumed stability is an illusion: even the most canonical texts remain subject to constant revision, and no two encounters with a text are quite the same. But in practice, this assumption remains crucial. When I put down my pen for the day, I rest in the belief that the same poem will await in the morning, that my notes will not magically vanish overnight. Those who study time-based arts such as music or drama therefore rely upon DVDs, MP3s, scores, scripts, and other artifactual media to enable sustained attention to their subjects. Taken on its own, improvisation opposes any such technology of preservation and repetition. The term stems from the Latin *improvīsus,* meaning "unforeseen," and it enters English in the late eighteenth century as a term of arts and oratory.[3] To improvise means to act *without foreseeing what one will do.* If improvisation refers to anything coherent, it means an absence of planning and predetermination. Yet sound recording stabilizes the event of improvisation; the moment of unplanned action gets captured on tape or disc. By making it possible to replay an improvisation, these technologies stabilize its precarious timing, its embrace of uncertainty, rendering it predictable and reproducible. As audio recordings circulate through archives, they are subsumed within processes of inscriptive capture and institutional control that run counter to the meaning of improvisation. We could not study improvisation without the technologies that preserve it, but improvisation as such opposes the kinds of predictability and repetition that recording enables. Indeed, many artists who value improvisation see it as freeing them from the rationalistic order and predetermination that any recording must involve. This elusiveness signifies improvisation's power as a mode of lyric exemption: it evades direct critical attention.

Scholars often discuss improvisation in terms of its others, a strategy of circumlocution. Paul F. Berliner, for instance, reports that musicians learn to improvise partly by memorizing a repertoire or "vocabulary" of short riffs and motifs they can recombine on the fly.[4] No doubt such preparations enable musicians to improvise well, but improvisation per se departs from and exceeds the ambit of any advance planning. Improvisation thus unfolds through a dynamic interplay with its others, and these others are often technological. Jazz improvisation became a popular technique concurrently with the maturation of audio recording technologies that rendered spontaneous performances endlessly reproducible.[5] Around the turn of the twentieth century, one of the first projects of the nascent record industry was to capture the sounds of Dixieland jazz and other improvised music. Half a century later, just as portable tape recorders became cheap enough for nonprofessional use, improvisation gained favor among the postwar avant-gardes. On one hand, we can learn a lot about improvisation by studying its technohistorical backgrounds. But on the other hand, this approach addresses improvisation in terms of its others, without directly confronting the poetics of the unforeseen, the undetermined.

By attempting not to reduce improvisation entirely to its technohistorical contexts, the present chapter dramatizes the difficulties that prevent more direct accounts of poetic improvisation from emerging—among them, the need for stable, repeatable objects of scholarly attention. To read the poetic record for improvisation also clarifies the reasons for poets' interest in it. By way of a brief example, these opening lines of a poem called "Dope," by Amiri Baraka, show how challenging it can be to read poetic improvisation:

uuuuuuuuuu
uuuuuuuuuu
uuuuuuuuuu uuu ray light morning fire lynch yet
            uuuuuuu, yester-pain in dreams

comes again. race-pain, people our people our people
everywhere ... yeh ... uuuuu, yeh
uuuuu. yeh
    our people
    yes people
    every people
    most people
    uuuuu, yeh uuuu, most people
    in pain
    yester-pain, and pain today
    (Screams) ooowow!   ooowow! It must be the devil
    (jumps up like a claw stuck him) oooo wow! oooowow!
        (screams)[6]

How might we discern improvisation in this text? The number of times "u"
appears in the opening lines seems arbitrary, so maybe Baraka improvises
these repetitions. But in the first line, "u" appears ten times, a suspiciously
round number, and the next lines begin with identical ten-letter bursts of
"u." The text seems carefully planned, rather than improvised. Reading a
similar Baraka poem, Meta Du Ewa Jones claims that such "strategic use
of typography and orthography" serves to "suggest improvisational techniques" directly in writing, without making a sound.[7] Even if the number of
repetitions varied arbitrarily, the written text would remain static, predictable, able to "suggest" improvisation but not to enact the spontaneity of a
live performance. We might also see traces of improvisation in deviations
between texts. The poem is formatted differently in the liner notes to *Poets
Read their Contemporary Poetry,* a 1980 Smithsonian Folkways record now
available for download: all lines there are flush left, not indented, and some
lines break differently.[8] Perhaps Baraka improvised these changes while
writing or rewriting, or perhaps they result from careless editing. Critics
hear improvisation in comparably small variations between recorded performances, but such attentive comparison cannot reliably distinguish improvisation from mere accident. Jones writes elsewhere, "The demands for
improvised jazz poetic forms might require that each performed iteration
of the poem be recognizably different from the previous one. But what
counts as difference?"[9] Improvisation begins when the performer's preparations get subsumed into the moment of spontaneous action, an indistinction between the planned and the unforeseen. When improvisation

leaves behind any record at all, it leaves traces of actions whose spontaneity cannot be disentangled from accident, contingency.

Despite this necessary indecision, obvious differences emerge between the printed text of "Dope" and Baraka's performance of it on the Folkways record. In this recording, he pronounces each "u" as the close back rounded vowel ("moon," "shoe"). Though "u" is printed ten times on the first line, Baraka pronounces it twenty-one times in quick succession before pausing, then twenty times while performing the second line, and then seventeen times for the third, with a shorter pause before the three "u" sounds that come after the space. Surely these divergences from the script count as improvisation. Indeed, a verbatim reading of the printed text would dampen its energy. Baraka does not pronounce "(Screams)" and other parentheticals but takes them as stage directions; he screams the words that follow. Perhaps the printing of "u" ten times likewise invites the performer to pronounce arbitrarily many "u" syllables in succession, a script that leaves space to improvise. Strangely, though, when he reads the shorter bursts of "u" later in the strophe, Baraka hews more closely to the indicated number of syllables. Like the printed text, his performance seems improvised in certain ways and carefully planned in others. Similarly, the recording itself offers rich evidence of improvisation while also sounding inadequate in important ways. For instance, we cannot say whether the poet "(jumps up like a claw stuck him)" because we have only a sound recording, not video. Moreover, the recording sounds the same every time you play it. This point may seem obvious, but it warrants careful attention. Sound recordings seem to capture the spontaneity of a live performance in a way that written language cannot, but their predictable repeatability provides crucial toeholds for critical attention, while also disrupting the illusion of spontaneity. Baraka pronounces "u" so quickly, for instance, that I could not accurately count the syllables, even after several listens, so I used sound-editing software to slow down the playback and obtain an accurate count—not the sort of thing one can do at a live performance. I will say more about this poem later, especially about how its racial politics and its textual conditions intersect with improvisation. For now, suffice it to say that reading improvisation in a printed text or a recorded performance presents significant challenges.

We might therefore view improvisation as a kind of limit concept. Its traces inhere in various kinds of records that make it available to sustained attention, but these records never fully capture the indecision, the live uncertainty, at the core of improvised action. Hence, when Fred Moten argues that "improvisation is never manifest as a kind of pure presence," he responds to the fact that it leaves only inadequate traces of itself. Moten theorizes improvisation as a dialectical energy that works through a determinate structure or "ensemble," counterposing artistic process and product.[10] The fleetingness of improvisation is one of the technique's main attractions; it signifies elusiveness, evading capture within the critical and technological faculties that stultify art by codifying and preserving it. But the very forms of capture that improvisation evades are the same that make it legible at all. The strain between improvisation's fleetingness and its openness to recording is its productive tension, from which its meanings emerge. Many artists value the radical spontaneity that improvisation enables, but to consider it truly immune to recording would impede the telling of its politically vital histories, specifically its links with African American artistic traditions. On one hand, then, an effective account of poetic improvisation must acknowledge the fleeting quality that constitutes much of its attraction. On the other hand, such an account must also address whether the valorization of spontaneous action engages with or turns away from the historical and technological contexts that situate it.

When electronics record poetic improvisation, a technique based upon uncertain futures and unforeseen actions, they not only register its politically salient histories but also figure its aesthetic elusiveness. In this chapter, I listen closely for improvisation in a few works by Baraka and his longtime friend Allen Ginsberg. Though both consider improvisation central to their poetics, they understand it differently. Baraka and Ginsberg remained friendly from their meeting in the late 1950s to Ginsberg's death in 1997, but their paths diverged in 1965 after the assassination of Malcolm X, when Baraka (then named LeRoi Jones) separated himself from the white avant-garde in Greenwich Village, moved to Harlem, founded the Black Arts movement, and eventually changed his name. The poets' theories of improvisation exemplify these different political commitments and poetic strategies. Despite these differences, the audio record of their works arrives through a shared network of technologies for recording, reproducing, distributing, preserving, and replaying poetic performance. Tracing these genealogies of recorded audio to the present, the chapter closes by discussing an online archive of recorded poetry that changes how we hear poetic improvisation.

The similarities and differences between Baraka and Ginsberg illuminate the political and formal stakes of poetic improvisation. Baraka acknowledged Ginsberg and other Beat writers as important early influences, but after the mid-1960s, when he distanced himself from the Beats' model of spontaneous creation, he also wrote less often about improvisation as a specifically black artistic tradition. Around the same time, however, his performance style became more improvisational. Like Ginsberg, by the 1970s Baraka had abandoned the clear diction and steady pacing that characterize both poets' early reading styles, and for the remainder of his career, Baraka would include singing, scat, shouting, and other improvised divergences from recitation of prewritten texts. During the same roughly five-year period, in other words, Baraka and Ginsberg altered their performance styles in strikingly similar ways, despite Baraka's separating himself from Ginsberg and the rest of the white avant-garde in the same period.[11] Many scholars view Baraka's move to Harlem as a decision against the poetics of spontaneous expression and in favor of a more organized, deliberated poetics of social justice. For these reasons, it seems inadequate to view Baraka's improvisations either as signs of affinity with white poets or as an essentially African American political tactic. His improvisations play both roles at once. By contrast, throughout his career Ginsberg more explicitly celebrated improvisation as a means of spontaneous creation, and his recorded performances more directly address questions about improvisation's technological conditions. They do not, however, illuminate the specifically political and historical dimensions of improvisation as powerfully as Baraka's.

Improvisation enables artists to resist the logic of cause and effect that produces foreknowledge and constrains our actions. Like other modes of lyric exemption, improvisation follows a two-step process by which, first, it is distanced from the order of rationalism and, second, it produces new techniques and objects that ground new critical insights, new knowledge. Improvisation's founding event is the moment of agentive free fall when we do not know what we will do next. To improvise means to act without deliberating in advance, and many value it for this distance it draws from rational planning.

Few studies of improvisation develop a positive vocabulary for this core idea of unplanned action. Most instead focus on the material, social, or generic contexts through which it unfolds, often claiming that such factors

hinder spontaneous action.[12] Such accounts see improvisation as constrained by the determinate structures whose order it would hope to exceed.
David Sterritt writes, "Spur-of-the-moment creation may not be nearly as
divorced from preconceived ideas, prerehearsed techniques, and prear-
ranged effects as its advocates frequently appear to believe."[13] Many such
critiques implicitly link practical spontaneity with political freedom, and
they forget that our social and technological environments support free
action even as they limit it.[14] Critics should be quick to recognize that literary invention, for instance, takes place through the structure of language,
which both constrains and enables new writing.[15] Those who minimize
improvisation's central precept, unplanned action, not only obscure its
main attraction but also ignore the challenges it puts to normal ways of
reading poetry. Improvisation offers a mode of exemption from the orderly,
predictable structures within which we think about poetry's historical and
technological dimensions. Before turning to Baraka and Ginsberg, I will
argue that these two contexts for improvisation, technological and historical, clarify how it challenges literary criticism.

Improvisation becomes available to sustained attention thanks to various
technologies of inscription, especially sound-recording devices.[16] Given its
basis in the unforeseen, improvisation seems to oppose the repetition and
reproduction that inscriptive technologies enable. Some claim that the very
notion of live performance emerges in response to phonographic recording, against which the live event gets defined.[17] This case is often overstated,
however, because live improvisation would already contrast with much earlier technologies of inscription such as the barrel organ, musical score, or
play script. The audience of improvisation comes together around an event
whose details are not predetermined and will not be precisely repeated.
The mere presence of a hot microphone might not stifle the spontaneous
action, but the status of the recording is less clear. Maybe recordings enable
us to hear the moment of improvisation again and again. It may sound
the same with every listen, but this does not change the fact that the performer improvised. Conversely, the recording's availability for endless
replay and reproduction might dampen the liveliness of improvisation.
Karl Coulthard writes, "Improvised jazz is always . . . new and unfamiliar
to an audience; however, a recording, even a live one, would seem to reify
this spontaneous process and turn it into a fixed text."[18] The most inventive performances sound rote after enough listens. Even when listening for

the first time, if we know an improvisation is prerecorded, it lacks the sense of unsettled possibility an audience perceives in live performance. Improvisation's tensions with inscriptive technologies also open its political meanings. Sound recording enables commodification; people can buy records instead of tickets to a performance. Recordings and musical scores have occasioned different legal debates about intellectual property, but both confirm inscription's power to reify musical performance.[19] Insomuch as improvisation evades the capture of recording, performers may see it as a way to resist this commodification in favor of the communalism of a live event. Recorded audio also has roots in state violence. Friedrich Kittler writes that "typewriting and sound recording are by-products of the American Civil War."[20] He notes that Edison's wartime experience as a telegrapher led him to develop the phonograph. Likewise, plastic audio tape was first developed in 1935 by the German firm BASF, which later made Zyklon B for the Nazi gas chambers. Basic sound recording predates Baraka and Ginsberg by several decades, but the equipment that makes their performances audible emerged from military research labs. Where improvisation escapes the capture of these technologies, perhaps it promises a way to oppose the deadly rationalism of the state. Improvisation evades the orderly technologies of information management that have facilitated the global spread of neoliberalism from the postwar years to now.[21] This tension between improvisation and inscription clarifies the meaning of the former, but it also complicates interpretation. If inscriptive technologies fail to capture improvisation, it becomes unclear how it can be available to the close scrutiny that scholarly interpretation normally demands.

Perhaps the most important contextual frame for improvisation is historical. Accounts of this type emphasize how improvisation emerged from African American musical practices, which respond to historical experiences of violence and oppression. If we discuss improvisation without addressing what William J. Harris calls "the African American audio past," then we further marginalize a historically oppressed people and, as I argue, ignore the very conditions that make improvisation attractive.[22] From this perspective, improvisation is not just an embrace of spontaneity; it developed from specific black musical traditions such as jazz, work songs, and spirituals. To theorize improvisation in the ahistorical present is to ignore the cultural memories of suffering to which its source genres respond. Perhaps improvisation's immunity to inscription refers not to some evanescent *now* but to the slave musicians for whom learning to write was punishable by lashing or death. Baraka and others make just such a claim. Improvisation might now provide a way to resist neoliberal informationalism, but the mothers and fathers of improvisation in America may have seen writing as a more immediate, bodily danger. Likewise, if improvisation embraces the precarious timing of uncertain futures, then perhaps this temporality figures other kinds of precariousness (political, economic, and social) that have attended African American life.

Given such considerations, it also becomes risky to celebrate improvisation for its opposition to rational planning. To praise an African American practice for its exemption from rationalism can read as primitivist or, worse, can support racist claims that people of color are irrational. This ambivalence recalls a difficulty seen in the first chapter, where the praise of affect clashes with the sexist coding of emotions as feminine, making it possible to derogate women as irrationally emotional. When African American artists celebrate improvisation for its difference from rationalism, they appropriate the terms of their oppression. This way around improvisation deemphasizes spontaneous action in order to attend the politically vital histories from which it emerges. The political claims made on behalf of improvisation often refer to its historical backings. If the immediate event of unplanned action forms the conceptual kernel of improvisation, though, it remains uncertain how political meanings can inhere in the practice without some predication of the context in which it takes place. In any case, this account indicates the dangers of theorizing improvisation without reference to the histories that have made it so influential.

PLANNING TO IMPROVISE: BARAKA'S PARATEXTS

No account of Amiri Baraka's politics or his performance style can ignore the change these underwent during the late 1960s. After service in the Air Force, Baraka arrived in Greenwich Village in 1957, still named LeRoi Jones. There he quickly befriended Ginsberg, Frank O'Hara, and other luminaries of the downtown poetry scene. He was soon publishing their work in his magazines, *Yugen* and *The Floating Bear* (the former coedited with his first wife, Hettie Cohen, the latter with Diane di Prima). Although Baraka acknowledges Ginsberg as an early influence, a rift soon opened between Baraka and the white avant-garde. After the assassination of Malcolm X in 1965, LeRoi Jones moved to Harlem, launched the Black Arts movement, and in 1967 changed his name.[23] His work thereafter exhibits an intensity of political commitment and racial consciousness largely absent from the poetry of his friends in the Village. During the same period of transition, his performance style shifted from staid academic diction to a more spontaneous and stylized delivery that involved singing, chanting, and shouting.

As we will see in the next section, Ginsberg's performances in this period likewise grew spontaneous and musical, but Baraka's more directly illuminate the political and historical meanings of improvisation. Harris alludes to these stakes when he compares two very different performances by Baraka, his August 1964 reading of "Black Dada Nihilismus" in Monterey, California, and his 1978 reading of "Dope" at Columbia University.[24] As Harris notes, Baraka's early "performance style drew heavily on white middle-class literary and verbal conventions," making his clear, measured diction "virtually indistinguishable from any other mainstream literary writer of the time." Baraka's later, improvisatory reading style, by contrast, "draws heavily on free jazz forms and traditions" that lead him "back to self, ethnic self, and ethnic history."[25] Through improvisation and other jazz-related techniques, Baraka invokes African American histories and identities. I have argued that someone who improvises does not let the past predetermine his next move; he prefers the open possibilities of an uncertain future. However, Baraka's improvisations call forth specific ethno-historical narratives about African American life. The electronic record of his performances makes it possible to explore how such politically salient narratives emerge from the traces of an aesthetic practice apparently concerned with evading historical determination.

Not everyone considers jazz the fundamental source of improvisation. Others associate it with oral communication, especially vernacular dialects and casual conversation. Jacques Coursil, for instance, links improvisation with "the *principle of non-premeditation of speech*," the idea that in ordinary contexts, "the chain of words" we speak "is not premeditated by the speaker."[26] The political implications of such accounts remain widely and intensely contested. But to equate improvisation with everyday talk detracts from its coherence. Earlier today I did not know in advance precisely what I would say to my colleague, to my partner, or to my dog. Perhaps I am improvising constantly! We all take countless unplanned actions every day. Indeed, our capacity to do so grounds our sense of agentive freedom, but to regard these little spontaneous acts as improvisation verges on absurdity. The most coherent theories of improvisation therefore address it as an artistic technique, taking jazz as their privileged example. This focus on jazz informs the political significance of improvisation. Because African Americans developed jazz, many read jazz improvisation as an allegory of African American historical experience or political commitments. As Nathaniel Mackey notes, the "African American improvisational legacy" has thereby "become a metaphor for all kinds of processes of cultural and social revaluation, cultural and social critique, cultural and social change."[27] Discussing Baraka in particular, Harris argues that he takes free jazz as "a contemporary way into African American history and tradition," a link to "the African American audio past." In particular, Harris claims that the free jazz influencing Baraka "is rooted in the shouts of the black church and the hollers of the field, sounds saturated with the history of slavery."[28] If improvisation means spontaneous action in the present, then this idea would seem troubled by its own historicity, but many consider the history of improvisation to be its most important dimension.

In turn, Baraka and others extend this historical reading to link improvisation with more current African American political stakes. As Jacques Attali puts it, for African Americans "the attainment of economic autonomy, the politicization of jazz, and the struggle for integration were contemporaneous."[29] Jazz improvisation thus functions at least as a metaphor for, and often as an instrument within, African American political struggle. Baraka argues as much in his 1963 study *Blues People: Negro Music in White America*, in which he claims that improvisation enables a "separation from, and anarchic disregard of, Western popular forms" of musical performance.[30] On one hand, the racially marked history of jazz grounds these political claims. On the other, improvisation's "anarchic" character, its basis in *not knowing* what comes next, informs what kinds of politics it enables. Walton M. Muyumba identifies improvisation as the "aesthetic practice best suited for negotiating the relationship between the anti-foundational, abstract self and the socially necessary black identity." If improvisation's historical origins in black music provide a metonym for "socially necessary" political struggle, in other words, then the unpredictability of improvisation can signify both the openness of revolutionary action and the "anti-foundational," universalist models of self from which people of color have been excluded and to which they might (or might not) lay claim. For this reason, as Muyumba argues, "in Jones/Baraka's

vocabulary 'blackness' is a synonym of 'improvisation.'"[31] No theory of
improvisation can afford to ignore the many ways "improvisation" in turn
functions as a synonym for "blackness." To read improvisation as a jazz
trope structures these ethnohistorical and ethnopolitical meanings.
As Baraka and others perceive, tensions emerge when we historicize a
practice whose vitality stems from its happening unpredictably in the present. In his 1959 manifesto "How You Sound??" Baraka embraces improvisation, declaring that "there must not be any preconceived notion or design
for what a poem ought to be." Avoiding premeditation also means deem-
phasizing historical precedents: "the only 'recognizable tradition' a poet
need follow is himself."[32] Baraka's techniques after his move to Harlem
heighten this tension between improvisation's liveness and its historical
dimensions; even as his performances became more improvisational, he
emphasized his work's continuity with the history of black artistic production. Mackey negotiates this tension by distinguishing between Baraka's
poems, which Mackey believes "aspire to the apparent conditionlessness
of the music," and Baraka's music criticism, which Mackey associates "with
the dialectical attachment of that conditionlessness to specific social, economic, and political conditions."[33] In similar fashion, as my readings of
Baraka show, the audio record of his performances brings the open conditionlessness of improvisation into contact with the determinate ethnic
histories from which such techniques emerged.

Many scholars view improvisation's historical specificity as limiting its
political powers, but these readings often neglect the core idea of unplanned
action. Muyumba, for instance, argues that "improvisation is a way of
articulating the agon among the individual artist, the musical setting, the
composition itself, and the larger social matrix that shapes the aesthetic."[34]
None of these, however, quite gets at the moment of unplanned action
that forms the conceptual kernel of improvisation. Similarly, Alexander
Weheliye argues that the "interplay between the ephemerality of music
(and/or the apparatus) and the materiality of the audio technologies/
practices (and/or music) provides the central, nonsublatable tension at the
core of sonic Afro-modernity."[35] Despite his deep commitment to the politics of "Afro-modernity," Weheliye says almost nothing about improvisation, likely because of his focus on recording technologies: he seems to
recognize the difficulty of reading the phonographic record for political
expressions that evade recording, that dissipate after the moment of live
performance. More pointedly, Melvin James Backstrom calls for skepticism

about linking improvisation with agentive freedom and political liberation. He identifies a fundamental opposition "between a wholly deterministic account of cultural production and the positing of its potential ethical-political significance," since the latter should involve some kind of free agency. He therefore doubts that improvisation can model true political liberation by operating "outside of the historically mediated structures in which it is implicated."[36] I do not share Backstrom's belief that politically salient action must be radically free of historical determinants, but he usefully clarifies the implicit relations among free agency, historical influence, and political liberation. In all three accounts, to interpret improvisation's histories limits the political significance of its central precept: unforeseen action. A contradiction emerges between improvisation's historical grounding in specific cultural practices and its conceptual grounding in undetermined action.

This contradiction is precisely the point. Jazz improvisation does not merely perform African American history but also figures that history as lost, impossible, effaced by slavery and its legacy. As Baraka points out, "the artifacts of African art and sculpture were consciously eradicated by slavery," an effort to debase and discourage the enslaved. Such violence made live performance more important because any art "based . . . on the production of an artifact, . . . such as a wooden statue or a woven cloth, had little chance of survival."[37] Improvised performances, by contrast, do not rely upon a script, and they leave nothing behind for oppressors to destroy. As a strategy for cultural resistance, improvisation finds strength in its fleetingness. Mackey links the "primacy of process" in Baraka's improvisatory poetics with a similar "bias against the artifactual," against art-as-product. Mackey associates this flexible, processual approach with an "anti-rationalist feeling" in Baraka's poems "that is also anti-Western" because it "implicates Western rationality in a range of exclusionary practices." The political salience of Baraka's improvisations, on such accounts, stems from the "dreamish, arational quality" that distinguishes them from rationalism's disciplinary work.[38] Others more directly link the politics of improvisation with the time of live performance. Coulthard notes that the pleasure of live music can "often be far more profound and lasting when there [is] no way to revisit the experience."[39] When he argues that "for most listeners unrecorded jazz is unknown jazz," he is not simply noting that the record industry helped to popularize jazz, nor that the genre's early years are lost to history. Rather, he constructs improvisation *as a lost object,* perceptible but fleeting from steady attention. He invests it with a pathos of imperfect recovery like that attending the Dickinson manuscript images in chapter 1. In both cases, the object's aesthetic interest lies in its ability to signify historically while also marking a site of historical loss. Lyrical thinking sustains such ambivalence in relation to historical knowledge throughout this study—in Dickinson's effaced but legible manuscripts, in Mac Low's chancy but deadly histories, and in O'Hara's address to an intimate but fleeting "you." In the case of Baraka's improvisations, this rhetoric of loss and recovery figures African American history not as an object of knowledge to be recovered but *as lost,* as the trace of an irrecoverable knowledge.

Improvisation thus unfolds a temporal structure particularly suited to figuring African American experience. To improvise means to construct the immediate future as uncertain, embracing the contingency of whatever happens next. Many see the precarious timing of improvisation as reminiscent of the broader political, social, and economic precariousness that African Americans have historically experienced. Mackey, for instance, hears in Baraka's poems a "fugitive spirit," a "fugitivity" that evokes not only the "mercuriality" of improvisation, its evasion of recording, but also a "relationship to enslavement and persecution" of African Americans, in response to which the fugitive voice "wants out," wants escape.[40] Likewise, Cornell West describes improvisation as "a spectacular form of risk-ridden execution" that "acknowledges radical contingency and even solicits challenge and danger" in order to figure the dangers of lived experience. West thus affiliates improvisation with "the highly precarious historical situations in which black people have found themselves."[41] Despite the contradiction between improvisation's historicity and its unfolding as an unplanned event in the present, its precarious timing figures the historical experiences of African Americans. The figural capacity of improvisation's timing, its ability to conjure uncertainty and vulnerability about the immediate future, may help to explain why African American artists find it such a productive technique.

The tensions around historically reading improvisation increase when the histories in question are local, personal. Mackey points out that improvisation means putting to use any available means and materials: "You take what comes to hand and you adapt it to your uses."[42] Any particular scene of improvisation, on this account, seems heavily determined by its immediate material-historical contexts. Many scholars also feel compelled to defend against the claim that, if a musician has spent years mastering his instrument and memorizing the licks he plays on the fly, then this cannot count as improvisation because he has planned and prepared so thoroughly.[43] Others take this conundrum as a sign that improvisation cannot happen. Michael Jarrett compares music to language, which we recombine to say new things: "The jazz musician is no different. He never actually improvises. He reworks what's given. Improvisation is an idea, one that's actually inconceivable."[44] Such arguments lead to the ridiculous conclusion that novelty cannot be produced in music (or language) at all, since everything is already given; they sidestep the really difficult work of describing how spontaneous actions and new meanings do in fact emerge from the given. Nonetheless, the model of improvisation as undetermined action becomes more difficult to sustain as closer attention is paid to the context of performance and the personal history of the performer. It becomes easier to attribute any detail of a performance to one contextual predetermination or another, implying that no part of the performance results from spontaneous choice.

Baraka's personal history likewise makes it difficult to read improvisation in his work. His early statements on poetics emphasized spontaneity, but this rhetoric diminished after his turn to Black Arts in the mid-1960s. Surprisingly, the immediate source of his early interest in improvisation was not jazz but Beat writing. His dawning racial consciousness did not lead him to reclaim improvisation as a Black Art, undoing white appropriation. Rather, as Daniel Belgrad argues, Baraka's decision to "reject the bohemian counterculture in favor of black power nationalism" coincided with his "becoming aware of the limitations of the spontaneous aesthetic."[45] Among these limitations is the fact that a poetics dedicated to the undetermined, spontaneous event cannot fully attend to its own cultural and historical conditions of possibility but must bracket these as inessential, lest they hamper the freedom of action. Baraka's early conviction that "the only 'recognizable tradition' a poet need follow is himself" may support a poetics of spontaneity, but it runs counter to his later emphasis upon cultural heritage.[46] One can see why a poet interested in social justice might abandon such a posture. To reclaim it on behalf of a black artistic tradition would seem politically ineffectual, if not contradictory. For Baraka, a more politically engaged approach to improvisation might deemphasize the event of spontaneous action and focus upon the historical contexts of

performance. To hear the most politically salient dimensions of improvisation, we might listen to the margins of poetic performance, asking how
the audible paratexts of a poem link it to its historical backings.

The term "paratext," borrowed from Gérard Genette, directs attention
to the peripheral and ambient noises that frame the recordings of Baraka's
performances, where the politics of improvisation become more audible.
Genette defines paratexts as the "accompanying productions" that appear
alongside a literary text, "such as an author's name, a title, a preface, illus-
trations." Other paratextual elements include editorial commentary, author
interviews, and reviews of the work. For Genette, a paratext "is always subordinate to 'its' text," serving to frame and structure attentions around it.[47]
By contrast, I share Moten's inclination to "think of the poem as the entire
field of saturation . . . within which the page, sound and meaning, the live,
the original, the recording, the score exist as icons or singular aspects of a
totality."[48] Yet, as Genette claims, the paratext refers to the text "precisely
in order to *present* it, in the usual sense of this verb but also in the strongest sense: to *make present*."[49] Without necessarily privileging written texts,
Baraka's paratextual performances *make present* the improvisatory dimensions of his work.

Genette divides the paratext into two categories, peritext and epitext,
and thereby clarifies the place of improvisation within the totality of a
poem. The peritext comprises any features within a physical publication
that do not count as the text proper, including page numbers, for example,
or copyright information. The epitext, meanwhile, "is any paratextual element not materially appended to the text," such as publishing contracts,
correspondence about the text, and "all public performances perhaps pre-
served on recordings."[50] Unlike the peritext, which authors and publishers
approve before publication, the public epitext is "circulating, as it were,
freely, in a virtually endless physical and social space."[51] The public epitext
thus remains unruly in relation to the text and opens possibilities for improvisation. As Genette notes, the epitext includes "interviews, conversations, and confidences, responsibility for which the author can always more
or less disclaim with denials of the type 'That's not exactly what I said'
or 'Those were off-the-cuff remarks.'"[52] Here, improvisation's significance
for the paratext almost gets lost in translation. For "off-the-cuff remarks,"
Genette's original reads, "*des propos improvisés*": improvised remarks.[53]
Developing at the margin of the poem, paratexts thus open an avenue for

improvisation to work upon and shape interpretations of it, especially when the paratext includes recordings of performances and other more or less spontaneous oral discourse.

An especially charged moment of improvisation can be heard at the start of the recording mentioned in the first pages of this chapter, Baraka's reading of "Dope" at Columbia University in 1978. Just as Genette understands a recorded reading as a paratext of a written work, the customary introduction of the poet by a host or emcee functions as a paratext to the reading proper. Before Baraka's performance begins, a male voice introduces the poet, but he flubs the introduction, mispronouncing Baraka's name. As the track begins, the emcee calls him "Mister Imaru . . ." and stutters, trails off, laughing briefly at his own blunder. He then tries again, offering "Emira Baraka. Okay?"[54] The final word sounds more sheepish than dismissive, but it does invite Baraka and the audience to accept the mispronunciation as close enough. The rhetorical question marks the poet's adopted Swahili name as linguistically and culturally alien, too strange to expect the emcee to get right. I cannot discover the identity of the emcee. As he interrupts himself with a self-deprecating chuckle and stutters through the unfamiliar words, the emcee signals the improvisatory character of this introduction, its status as an unplanned paratext of the scripted poem about to be read. Earlier, we identified a tension between the moment of spontaneous action and the cultural contexts enframing it, but here the encounter with cultural alterity directly occasions improvisation. If the emcee spoke Swahili, or if Baraka had a more Anglophonic name, the introduction would likely have gone smoothly, or at least a flubbed introduction would not sound politically resonant. In the paratext of poetic performance, the opposition between the improvisatory event and its enabling contexts seems to break down. As the emcee excuses himself, the audience applauds Baraka. Taking to the microphone, the poet begins by saying, "I'll be alright." This is a strikingly effective, improvised response to the emcee's error. It assures everyone that Baraka does not feel threatened or insulted by the poor introduction, but this assurance is subtly barbed. It implies that the emcee did not, in fact, manage a correct pronunciation of the poet's name. It also draws a contrast between the poet who is "alright," poised in this moment of improvisation before he reads, and the flummoxed emcee who, perhaps, is not alright because he has embarrassed himself.

The opening pages of this chapter suggest one can read improvisation in "Dope" by tracking the differences between the written text and Baraka's oral performance, or even between written versions of it. But this strategy makes improvisation evasive of critical attention, lost in the play of intertextual differences. Other critics take a stylistic approach, arguing that Baraka's oral and orthographic techniques invoke the jazz tradition of improvisation without literally improvising. Yet a third way of reading improvisation becomes available in the opening seconds of the track. Improvisation seems most available to direct attention in the paratextual moments of a recording. So much that sounds improvised in a performance falls outside the speaking of the poem itself: it is the poet's quip before he begins, his cough or stutter as he reads, his offhand comments between poems. Even involuntary noises remind us that improvisation functions as an index of free agency in general, since improvised actions resist predetermination. Meta Du Ewa Jones productively associates such peripheral moments with "jazz improvisational standards" of performance, but her decision to call them "extra-textual vocalizations" has the unhappy effect of recentering the written text as a script for performance.[55] Although such noises certainly remain exterior to the alphabetic poem, they form a crucial part of the recorded poem as audio-text, a paratextual dimension of the recording that helps to mark its difference from alphabetic writing.

As recorded poetry readings become more widely accessible online, it is not a given that such paratextual moments will remain audible. Harris's excellent interpretation of the 1978 performance of "Dope" at Columbia, for example, does not mention Baraka's strange interaction with the emcee.[56] If we can trust Harris's list of references, it seems not that he ignores this exchange but that he never hears it. His essay cites an album that truncates the first twenty-seven seconds of the track, omitting the flubbed introduction, Baraka's response, and his prefatory comments about "Dope."[57] Both versions of the track remain available on the Smithsonian Folkways website, but the site does not indicate their differences, nor even the fact that they are recordings of the same event. The current online infrastructures for archiving recorded poetry thus seem both too messy, lacking reliable systems of reference, and also somehow too tidy, since they often elide paratextual details. Future archives of recorded poetry face the dual challenge of organizing their contents in a systematic, accessible manner while also seeking to preserve as much of the audio record as possible.

In the printed text of "Dope," Baraka launches an attack upon a major pillar of African American musical culture: the black church. His critique against the politics that emerge from religious faith accords with his turn to Marxism in the mid-1970s. It also resonates with similar critiques he had already deployed against Ginsberg's approach to spontaneity. In both cases, he objects to religion that detracts from real political concerns. In "Dope," he excoriates black communities for attributing "race-pain" to "the devil" instead of the rich.[58] Before reading the poem in July 1978 at Naropa University, Baraka mentions the people in Newark "sticking dope in their arm" and describes the text as "a poem about a different kind of dope being stuck in peoples' heads."[59] In this poem, the people of Newark repeatedly insist that "it must be the devil" causing their troubles.[60] Preferring faith over political confrontation, they insist that it "cain be rockefeller" keeping them down:

> caint be him, no lawd
> aint be dupont, no lawd, cain be, no lawd, no way
> noway, naw saw, no way jose—cain be them rich folks
> theys good to us theys good to us theys good to us theys
> good to us theys good to us, i know, the massa tolt me
> so, i seed it on channel 7, i seed it on channel 9 . . .[61]

Ventriloquizing the fast, desperate jive of a junkie, Baraka criticizes African Americans for embracing the narcotic of religious faith. He figures the mass media as a postmodern slave master who convinces those in the ghetto not to blame the rich. He also indicts the black church for instilling a quietist faith in the kingdom of heaven:

> must be the devil, going to heaven after i die, after we die
> everything gonna be different, after we die we aint gon be
> hungry, ain gon be pain, ain gon be sufferin wont go thru this
> again, after we die, after we die owooo! owowoooo![62]

The unconventional spelling and onomatopoeias provide leeway to improvise when performing this poem; the whoops and cries of the poet resonate with the enthusiasms of the same black church his poem lampoons. Although Baraka had initially separated himself from the white avant-garde in order to advance a black nationalist agenda, he soon found reason to attack black false consciousness on the same grounds from which he criticized his white literary peers. The African American spiritualism that keeps faith in redemption "after we die" while avoiding conflict in the present seems equally guilty of preventing real political progress. This antireligious framework for improvisation rejects the spiritualism some associate with the practice. Moten, for instance, claims that improvisation "always also operates as a kind of foreshadowing, if not prophetic, description" that reverses the word's etymology through techniques of foresight.[63] Baraka would have resisted the framing of improvisation as prophecy just as decisively as he rejected the mysticism of white poets interested in spontaneity.

Baraka's most pointed criticism of Ginsberg, then, does not concern the Beat writers' whitewashing of improvisation. Instead, he attacks Ginsberg on the same grounds from which he would later criticize African American religious practices. In a poem from the late 1960s, "Western Front," he derides Ginsberg for his mysticism and detachment:

> Poems are made by fools like Allen Ginsberg, who loves God, and went to India only to see God, finding him walking barefoot in the street, blood sickness and hysteria, yet only God touched this poet, who has no use for the world. But only God, who is sole dope manufacturer of the universe, and is responsible for ease and logic. Only God, the baldhead faggot, is clearly responsible, not, for definite, no cats we know.[64]

Baraka counts Ginsberg among the "fools" who write poems, an extension of the ancient trope that casts poets as irrational. Ginsberg embraces this trope, but for Baraka, it signifies a pernicious obliviousness. Just as "dope" in the poem by that name means both drugs and religion, so in this poem, the "dope" that God manufactures may be faith, drugs, or dopes like Ginsberg. Alluding to Ginsberg's 1962–63 journey through India, Baraka claims that Ginsberg's desire "only to see God" leads him to ignore the realities of "blood sickness and hysteria" among the poor. Baraka calls God a "baldhead faggot" to index Ginsberg's narcissistic spiritualism, his finding "no use for the world," no concern about real living conditions. The poem ends with a sarcastic ventriloquism of Ginsberg's view that the suffering of the poor must be caused by God, "not, for definite, no cats we know." To be fair, Ginsberg did undertake significant political action in the late 1960s and after, especially against the Vietnam War and nuclear armament. As we will see below, however, it is easy to recognize Ginsberg's spiritualism as a distraction from real human oppressors. Baraka implies that "cats we know" must be held accountable.

Although both Ginsberg and Baraka embrace improvisatory performance styles in the late 1960s, Baraka would link his with African American
traditions while dismissing Ginsberg as politically oblivious. The parallel-
ism of Baraka's two antireligious critiques, however, would eventually help
to facilitate rapprochement between the poets. Ginsberg and Baraka were
together again for Baraka's 1978 reading at Naropa University. Introducing
his friend, Ginsberg recounts "a long period of paranoia and separation."
He blames "government agents" for persecuting Baraka and keeping him
"isolated from some of his earlier friends."[65] Ginsberg's voice cracks with
emotion as he says, "I'm really happy that he's here with us now. There's
some kind of reconciliation." As the rustling of papers suggests, this introduction is prewritten or extemporized from notes. But the moments
when Ginsberg's voice slows, softens, and cracks with emotion—these
seem unplanned. In fact their apparent spontaneity helps to guarantee the
authenticity of Ginsberg's feelings. On the other hand, those parts of the
recording that do sound scripted betray the lingering tensions between
these poets, as when Ginsberg studiously blames their rift on government
agents rather than Baraka's turn to racial politics. Baraka starts his performance by reading a "brief statement" about art and "the class struggle
within society." At this event, making a political statement or even recount-
ing a history appears to mean reciting a written script, while moments that
sound improvised seem less politically salient.

Despite the common association between African American histories
and improvisatory practices, it is not clear that Baraka viewed improvisation as doing political work in his post-1965 performances or that his
major political commitments directly concerned improvisation. His opening statement at the 1978 Naropa reading sets up an opposition between
improvisation and the organized, planned actions he considered more
efficacious. "In the '60s," he recalls, "without a genuine Communist Party,
the movement was left to spontaneity." Because these political energies
lacked an organizing structure, they ultimately "sunk into cultural nationalism, black capitalism, gun cultism, reformism, metaphysics, idealism."
Before 1965, Baraka viewed improvisation as a temporal figure through
which black artists respond to experiences of racial oppression and precariousness. His poetry readings in this period, however, remained staid
and academic. After his turn to Black Nationalism in the late 1960s, he
viewed "spontaneity" as an insufficiently organized mode of political action.

Nonetheless, his readings in the late 1960s, and those for the rest of his career, would involve an array of improvisatory techniques. For some listeners, the stakes of such improvisations may seem stylistic or aesthetic, rather than immediately political. His 1978 reading of "Dope" at Naropa University begins with the poet rapidly pronouncing "u" as the rounded vowel, just as he does in the other 1978 recording of this poem, discussed earlier. At Naropa, though, he pronounces "u" thirteen times for the first line, twelve for the second, and fourteen for the third. As noted above, the other recording of "Dope" has Baraka pronouncing "u" twenty-one times for the first line, twenty for the second, and seventeen for the third. These performances differ from each other and from the printed poem, where "u" appears ten times in the first line, ten in the second, and thirteen in the third. The written poem structures his reading, therefore, but it does so by providing a normative script from which improvised performances diverge. Likewise, Baraka shouts the exclamations, "Owooo! owowoooo!" cited above, but they sound different depending on the timber and volume of the poet's voice, the recording equipment in use, and the acoustics of the performance space. Meta Du Ewa Jones argues that Baraka's improvisations rely on a "sense of immediacy" and of live action that "is illusive, since these performances are mediated through the realms of videotape and vinyl," which make them predetermined and repeatable.[66] Despite emphasizing Baraka's orthographic techniques, Jones admits that "the performance environment does allow for flexibility in the form and tone of his verse unrealizable in the printed version." This sense of flexibility, of spontaneous deviance from the written poem, persists in recorded versions of it. Audio recording registers and preserves differences between performances, making it possible to read improvisation in these variances. As paratexts to the written poems, they serve "to *make present*" (per Genette) the improvisatory dimensions of Baraka's practice.[67]

Especially in the later decades of Baraka's life, his performances include improvisations that lack obvious correlates on paper. In "Wailers," his tribute to Larry Neal and Bob Marley, Baraka asks, "You wanna imitate this? / Listen. Spree dee deet sprree deee whee spredeee shee deee // My calling card."[68] When he performs the poem, he makes noises that significantly diverge from those the written onomatopoeias suggest. In the 1982 film *Poetry in Motion* (directed by Ron Mann), he reads the piece accompanied by a jazz saxophonist and drummer, and the onomatopoeic line sounds like the saxophone.[69] Although the initial "Spree" syllable can be heard here, the sequence ends with an "ah" sound not represented in the written poem. In another performance of "Wailers," recorded at UCLA on May 18, 1983, Baraka pronounces the first few words more or less faithfully, but his line of scat ends with shouts of "wow!" to which the audience applauds.[70] He performed the poem again at Naropa University in 1984, and there the onomatopoeic sequence sounds more like the scramble of a fast-forwarding audio tape.[71] The written "Spree" syllables are audible, but an unwritten "spdoo" can also be heard. These improvisations differ not just quantitatively, as with the "u" sounds of "Dope," but also qualitatively. Baraka refers to this line of improvised scat as his "calling card," and he suggests that any attempt to "imitate this" will not yield successful reproduction but only more spontaneous divergences from the script. The written scat line provides a norm from which to diverge, a microcosm of the "changing same," Baraka's term for the historical continuity of African American culture.[72] Its coherence across the centuries stems from the embrace of change, of divergence, that motivates African American improvisatory practices. In Baraka's performances of "Wailers," these paratextual divergences from the script are audible not because it is impossible to write down the sounds Baraka makes. One can easily develop onomatopoeias to approximate the sounds of each performance, as I just did. And the tape recorder directly inscribes the scat line as Baraka performs it. Hence, the paratextual divergence of this line, its disruptive adjacency to inscription, stems not from the limitations of writing but from the poet's spontaneous invention of new scat sounds at the time of each performance.

In some of the latest recordings of Baraka available online, the poet sings or hums, often between poems. This technique again positions music as paratextual in relation to the written poem. Less clear, however, is the possible status of the singing as improvisation. On one hand, moments of song sound casually interspersed between the poems, loose melodies that set the mood. On the other hand, close listening suggests that Baraka's singing is planned and executed in accordance with a script. For a 1992 reading at Naropa University, Baraka opened with another "statement" on politics and then performed the poem "Masked Angel Costume."[73] He sings a few notes between each short, numbered section of the poem. These brief musical interludes sound vaguely familiar and have melodic affinities with one another, but Baraka sings them so casually that one might assume he improvises. The final section of the poem, however, mentions "*Birmingham* / ... where / 4 of my daughters / were killed," a reference to the 1963 bombing of the 16th Street Baptist Church, which killed four African American girls.[74] The next strophe recalls how "John Coltrane / composed / Alabama," a song written in response to the bombing. From the start of the poem, it turns out, Baraka has been humming bits of the saxophone line from the Coltrane song. The poem does offer an early clue about the source of Baraka's song, for it includes a strange headnote. Here is how the title, headnote, and opening lines appear in *SOS*, Baraka's volume of collected poems:

MASKED ANGEL COSTUME

THE SAYINGS OF MANTAN MORELAND
Alabama

1   Never let a ghost
    Ketch you
                Never!

2   Avoid Death[75]

Before reading the poem at Naropa in 1992, Baraka briefly mentions Mantan Moreland, a black actor best known for his comedic work in the 1920s and 1930s. Baraka then says, "This is called 'Masked Angel Costume: The Sayings of Mantan Moreland.'" He does not pronounce the next word of the printed poem, "*Alabama*," but instead says, "one," sings a few bars of the Coltrane song, and continues with the script, "Never let . . ." He sings more Coltrane before saying, "two," and goes on from there. His performance disrupts the normal relation between a written text and its oral recitation. The italicized word "*Alabama*" is not simply a stage direction but a hermetic paratext legible to Baraka alone. A reader who encounters this poem in print, without hearing Baraka perform it, probably will not recognize "*Alabama*" as a performance note and will read the word aloud. Even if the italics do suggest the word is paratextual, a first-time reader will have no way of knowing it refers to a Coltrane song until she has reached the end of the poem. Even a reader who has read the poem before and knows the Coltrane tune will have no way of knowing *when* in the performance of the poem to sing. After all, Baraka does not sing where the song title appears in the text but only after reading "one," and he sings again between sections without any textual direction to do so. For Baraka alone, the single word "*Alabama*" means "sing bits of Coltrane's 'Alabama' before reading each section of the poem." Baraka's paratextual singing, therefore, disrupts the model of poetic text as a script that can be adequately performed by anyone who speaks the language. This text does not include sufficient information to perform it as Baraka does. The singing is scripted but only loosely so, and in a way only Baraka can read.

Baraka's paratextual singing challenges the assumption that print accurately conveys language to an audience temporally or spatially distant from the author. Considering the printed poem apart from Baraka's recorded performance, the italicized "*Alabama*" appears as the inert remainder of the poet's intention to riff on Coltrane. The word persists on the page without the power to signify as it does in Baraka's hands. Only upon hearing Baraka's performance does the poem resonate with Coltrane's response to the murder of four little girls in Birmingham. The paratextual situation of the singing alongside the reading makes it unclear where the scripted performance might end and the improvisation begin. Likewise, Baraka's casual riffing on a politically salient jazz song makes it difficult to say whether he means us to hear the entire poem against the backdrop of Coltrane's protest and mourning or to make the connection only in retrospect, when he concludes the poem by mentioning Birmingham and "Alabama" directly.

Other poems in this 1992 performance invoke similar paratexts for singing. In print, the poem titled "Why It's Quiet in Some Churches" begins with an italicized headnote, "*Just a Closeta Walk with Thee,*" Baraka's vernacular spelling of the old gospel song "Just a Closer Walk with Thee." Instead of reading the headnote aloud, Baraka announces the title of the poem, sings a few lines from the song, and then recites the poem. Within a few lines, the poem suggests that homophonic play of the sort Baraka deploys in the headnote can have political significance: "We changed the spelling of Prophet to Profit."[76] This joke is easy enough to grasp without seeing the printed text, and in fact the effect is greater when the listener must negotiate the words' phonic similarity without seeing their orthographic difference. The audience at the reading, however, cannot see that a similar homophonic figure appears in the vernacular spelling of the headnote, since Baraka does not read the note. Instead, before he starts to sing he briefly offers, "the wealthier the church, I'm talking about black churches, the more European," which primes his audience to hear his gritty gospel singing as a gesture toward "African" roots. Similar messages about the political unruliness of traditionally black sounds emerge from the printed and the performed versions of this poem, then, but the message emerges differently because of the gospel paratext's different relations to the visual and aural dimensions of the poem. Later in the same reading, Baraka recites "Funk Lore," which includes a headnote, *"Blue Monk."* Instead of speaking these words, he sings some bars of the Thelonious Monk song by that name. Likewise, the poem "I Am" bears the headnote *"Blues March,"* and again Baraka sings the Art Blakey tune by that name instead of speaking the printed words.[77] In these three cases, Baraka simply prefaces each poem with a little singing instead of interspersing song mid-poem, but here too there is a complex relation among the printed poem, its oral recitation, and its musical paratext. Baraka reproduces the melody of each song with impressive accuracy while still singing with a casual, offhand tone that suggests improvisation. As paratexts to the poems, these bits of song bring Baraka's work into contact with a tradition of black musical improvisation whose cultural sources and political motives his own works extend.

Recordings of Baraka's performances thus productively dramatize this tension involved in thinking through the political significance of improvisation. Over the course of Baraka's career, improvisation became an important performance technique for him. His strongest claims about improvisation's ethnic and political meanings came early, when his own reading style remained quite traditional, but his actual practices of improvisation continued to develop late into his life, as did his understanding of its political significance. As we will see in the next section, some writers, including the Beats who influenced Baraka's early work, associate improvisation with a politics of radical freedom. Others, including Baraka, emphasize not only improvisation's open-ended, eventual structure but also its history as an African American artistic practice. Reading these histories of improvisation contributes to a politics of racial justice in at least two ways. First, because improvisation involves the precarious timing of not knowing what comes next, it provides a temporal figure for the kinds of precariousness and vulnerability that have long textured African American life. By exhibiting poise and mastery in a time of uncertainty, the improvising black artist reenacts the ancestral, improvisatory techniques of survival. Second, because the moment of improvisation is fleeting, the technique constructs the past as a lost object, never fully available to critical attention, and this figure of lost histories recalls the oppressive campaigns of cultural erasure that African American communities have often endured.[78]

Despite these political affordances, Baraka recognizes that to emphasize the history of improvisation is to tarry with a latent contradiction. On one hand, in jazz and other arenas, improvisation is rightly associated with the racial identities and cultural histories of African Americans. On the other hand, improvisation involves a moment of action in the present that is not predetermined, not shaped and constrained historically, so emphasizing its historical dimension would seem to mean vacating its central precept. Baraka's performances embrace both the evental structure of improvised action and the politically salient historical dimension of improvisation, its links to the African American past. One can read improvisation not only in the differences between written and spoken versions of his poems but also in the paratextual elements of performance, where his improvisations sound most politically engaged and most closely in touch with the history of black improvisation. Whether Baraka sings a few measures of jazz or makes a wry comment to his host, his improvisations bear the live political energies of spontaneity while also extending and reworking the long history of improvisation as a black artistic technique.

IT JUST CLICKS: GINSBERG ON TAPE

Having addressed poetic improvisation's historical meanings, we can now more directly explore its technological dimensions. To do so, this section reads Allen Ginsberg's poems, interviews, and performances from the late 1960s and beyond. Multiple scholars have addressed Ginsberg's performance style, but most focus on his earlier readings, especially his debut performance of "Howl" at Six Gallery, San Francisco, on October 7, 1955. This was a watershed event for American literature, but the reading itself is not easy to study, because it was not recorded and quickly became fodder for literary mythologizing. Recordings from the same period reveal that Ginsberg had already begun to adopt a less academic, more energetic, incantatory performance style, but it was not until the 1960s that his performances began to include chanting and more overtly spontaneous techniques. Like Baraka, in this period Ginsberg developed a more improvisatory approach to performance, abandoning the more conventional ways of reading aloud that continued to structure his recitals in the 1950s. Unlike his estranged friend, however, Ginsberg also emphasized technology's role in his creative practice, thereby enabling us to ask how electronic media, and especially audio recording devices, intersect with poetic improvisation. Ginsberg saw electronics as shaping his sense of poetic form and of the political work his poetry can do. His view was ambivalent, however, as electronics both enable and constrain the formal and political potentials he saw in improvisation.

By the late 1960s, Ginsberg had become a public representative of the counterculture. He had frequent occasion not only to read and discuss his poetry before large audiences but also to talk about the social contexts in which his poems participated. In a May 1968 interview on William F. Buckley's television talk show *Firing Line*, Buckley introduces Ginsberg as "the hippies' hippie, the Bohemian prototype."[79] Ginsberg soon suggests that "hippie" is a "newspaper stereotype," and this point leads him to a sharper attack on the constraints Buckley's program placed upon him:

> The problem, I think, with television, which is interesting—see, before I came on, uh, one of your, uh, Mr. Steibel who is your, uh, uh, helper here, producer, asked me not to say any dirty words, what he considered dirty words, on the program, which presents a moral problem, you know, in that, um, there's a political function to the language of everyday use—the language we actually speak to each other off the air. There's a communication that's involved, and there's a classical use of all sorts of what are called off-color words in art, as well as images. So our problem here, or what I've been posed with, is having in a sense to, uh, censor my thought patterns.

Buckley responds that Ginsberg's commitment to "love" must entail consideration of others: "if it would be offensive to some people to hear those words," then Ginsberg should "assume the burden" of avoiding offense. Ginsberg defends himself in two ways. First he claims that he "could probably pronounce" off-color words "in such a way that would be inoffensive, and in fact everybody would shout with joy on the television screen watching." Indeed, Ginsberg's ecstatic reframing of the vulgar counts among his strengths as a poet. His second justification returns to the politics of televised speech:

> See, one of the problems is that, for instance, uh, politically speaking, no one can understand the problem of police brutality in America, or the police state scene that we're going through, as I see it, without understanding the language of the police, but the language that the police use on, say, hippies or Negroes is such that I can't pronounce it to the middle-class audience.

So the middle-class audience doesn't really have the actual data, or some portion of the data, to judge the emotional situation between the Negroes and the police.

Ginsberg frames the control of language on television as a political and moral problem. It limits his "thought patterns," forecloses artistically valuable uses of profanity, and constrains political discourse. Censorship had concerned Ginsberg at least since the 1957 obscenity trial connected with his first book, *Howl and Other Poems*. More recently, in January 1965, he testified at the Boston obscenity trial of William S. Burroughs's *Naked Lunch* (1959), where he praised the book's treatment of the "addiction to controlling other people by having power over them."[80] Facing Buckley, he voices concerns about similar control over speech on electronic media. Censorship, he argues, can impede what many consider the primary function of such media: the broadcast of information to mass audiences. Regulations against profanity on television keep politically salient "data" from reaching viewers.[81] In this sense, television deprives us of information instead of providing it, and thereby accomplishes a sort of mind control. Nonetheless, Ginsberg also approaches television and other electronic media as powerful means to disseminate his poetry and his political messages. Lesley Wheeler sees Ginsberg's performances as "representing the opposite of the polished, distant televised world," but his attitudes about such media are more complex than Wheeler recognizes.[82] These contexts matter for a reading of Ginsberg's improvisatory practices because they specify how the poet believes electronics enable and constrain his language, especially his spontaneous performances.

Despite these limitations, Ginsberg considers electronics worthy of serious attention in his poetry. His writing from the 1960s offers various political and metaphysical figures for electronics. A poem in his book *Planet News* (1968) confesses, "now I'm paranoiac every day about the cops / ... as if it were all being tape recorded from my / skull ... / He saw electric wires on the floor—He saw / the channel that heard yr mind."[83] As on *Firing Line*, electronics serve here as a medium of thought and a technology of government control. Later in the same book, "A Vision in Hollywood" culminates in a kind of electric ontology, "with moon lights, neon arc-flames, electric switches," and other strange technological imagery. The mystical messages of these poems seem to arrive through electronic figures. Such media also seem to offer the solace of human solidarity, as when "all telephones ringing at once connect every mind by its ears to one vast consciousness."[84] In this book, electronics help Ginsberg to imagine a transcendent collectivity. Another poem, this time in *The Fall of America* (1972), describes political debates on talk radio as "electric arguments / [that] ray over / the ground." He even cites these arguments directly: "'... I dont see any reason' says the radio / 'for those agitators— / Why dont they move in with the negroes?'" Electronics may hold the promise of collectivity, but they also heighten political tensions as we listen to what "the radio" says, not to other people. "Radio" may be "the soul of the nation," but racism and divisiveness evidently persist.[85]

The Fall of America also deploys electronic tropes on artistic production, again making it possible to discern how Ginsberg thinks these technologies enable and constrain his improvised poetics. The opening page describes "Bob Dylan's voice on airways" not as an intimate trace of the singer but as a "mass machine-made folksong."[86] By the mid-1960s, neither Ginsberg nor Dylan viewed electronics simply as instruments of authority to be opposed with a folksy primitivism; they instead took such technologies as means to make art. The summer of 1965 saw the "electric Dylan controversy," as Dylan's turn to an electric rock sound proved divisive among his fans. Many worried he had sold out. Coming to his defense, Ginsberg declared: "Dylan has sold out to God. That is, his command was to spread his beauty as wide as possible," and electronics helped him to do so.[87] Ginsberg, for his part, used electronics to compose many poems in the 1960s. Instead of beginning with a typewriter or a pen, he spoke extemporaneously into a tape recorder. A few months after Ginsberg wrote about Dylan's "machine-made folksong," the singer gave him money to buy a portable Uher tape recorder, which Ginsberg carried on a cross-country road trip. While others drove, Ginsberg spoke poems into the machine, among them the famous antiwar poem "Wichita Vortex Sutra." Electronics did not simply provide Ginsberg with a tropology for political diagnosis or a vocabulary for spiritual aspirations; from the tape recorder to the television camera, electronics also enabled him to produce and disseminate poetry.

In a 1968 interview titled "Improvised Poetics," Ginsberg describes how his use of audio tape affects the formal features of "Wichita Vortex Sutra" and related poems. Although the tape recorder helps to capture his spontaneous language more directly, its technical features also influence the shape a poem takes, exerting a determinative force against the purported freedoms of improvisation. Meanwhile, the printed form of this interview transcript itself illuminates the limits of inscription as a means to preserve improvised language. The relevant discussion begins when the interviewer, Michael Aldrich, asks Ginsberg why he indents a few lines of "Wichita Vortex Sutra" in a certain way. Ginsberg does not exactly answer the question, but he describes his process:

> The way this was determined was: I dictated it on this Uher tape recorder. Now this Uher microphone has a little on-off gadget here (click!) and then when you hear the click it starts it again, so the way I was doing it was this (click!); when I clicked it on again it meant I had something to say. (*SM*, 134)

As an interview transcript, this account opens more questions than it answers. Does Ginsberg orally pronounce the "(click!)" each time it appears, or does he demonstrate the noise on the tape recorder itself, as "this" and "here" suggest? If he demonstrates on an actual tape recorder, is it the same device being used to record the interview? If so, then Ginsberg's strategy for explaining the machine's function leads him intermittently to deactivate the very device recording his explanation. Perhaps he said something while the machine was turned off. As we read the printed transcript of the interview, we cannot know. Talking about the tape leads directly to the limits of its function. Alternatively, if Ginsberg had demonstrated on a machine other than the one recording the interview, then one wonders how the sound of one recorder switching on, as captured by a separate recorder, differs from the same sound as rendered internally by the machine that switches on. The onomatopoeic "(click!)" flattens this possible difference on the page, rendering it uncertain. These seem like picky questions about a brief moment in one interview, but they indicate the fundamental difficulty of accurately tracking the interplay among improvisation, audio recording, and alphabetic writing.

These difficulties intensify moments later in the same interview when Ginsberg demonstrates how this "(click!)" develops poetic form. To do so, he recites the lines of "Wichita Vortex Sutra" just cited by the interviewer, but with a curious addition:

> That the rest of earth is unseen, (Click!)
> an outer universe invisible, (Click!)

Unknown (Click!) except thru
        (Click!) language
           (Click!) airprint
              (Click!) magic images
or prophecy of the secret (Click!)
    heart the same (Click!)
    in Waterville as Saigon one human form (Click!) (*SM*, 134)

No other printing of "Wichita Vortex Sutra" or any related poem includes the "(Click!)" that indicates where Ginsberg pauses the recorder during his oral improvisation of the poem. His addition of the "(Click!)" does elucidate a formal principle, but in other ways it perplexes. Most straight-forward, the passage demonstrates that Ginsberg breaks lines wherever the click of the audio tape indicates he paused and then resumed recording to speak the next phrase. This rule of thumb does not, however, answer the interviewer's original question about indentations. He later explains the indentations as a means to "arrange the phrasings on the page visually, as somewhat the equivalent of how they arrive in the mind and how they're vocalized on the tape recorder," but this account remains unclear (*SM*, 135). The length of an audio tape physically represents the single dimension of time along which language is spoken. Ginsberg fails to say how the second dimension of a page, its width, across which he indents lines, contributes to his written presentation of speech. Drawing upon the breath-based prosody of William Carlos Williams, he apparently bases formal elements of the printed poem on "natural speech pauses," where he stops the tape, which in turn "indicate mind-breaks," the poet's term for interruptions of thought (*SM*, 126). But uncertainties about "(Click!)" persist. Does Ginsberg orally pronounce the word here or use a tape recorder to make it? Why does the "(Click!)" appear at the end of some lines but at the start of others? In one case, it appears mid-line without the expected line break, but the poet gives no explanation for this inconsistency. And why is "Click" capitalized in the verse but not in the prose cited above? Ginsberg's verse lines do not begin with capitals, so it is not a poetic convention. Perhaps the clicks become louder when Ginsberg recites his poetry, or perhaps the transcription is simply inconsistent. Although the poet leaves us ill equipped to answer these questions, the click of his tape recorder provides an integral formal texture for his improvised poetic discourse without strictly determining its written form.

Ginsberg describes the print versions of "Wichita Vortex Sutra" and related poems as "transcribing" his recorded speech, but their form marks the unreliability of such transcription (*SM*, 135). Why do the published versions not include the "(Click!)" as it appears in the interview transcript? Certainly they could. The parenthetical exclamations give each line a satisfying closure. The ongoing clicks evoke the finger snapping associated with jazz performances and Beat poetry readings. They also presage the chanting that would increasingly interest Ginsberg in the late 1960s and beyond. Although Ginsberg understands his line breaks as formal equivalents of the audible clicks of the tape recorder, this visual presentation of each "(Click!)" on the page could appear as a more direct transcription of the recording process. Certainly it gives a clearer sense of how the technical function of the tape recorder guides his poetic form and aids his improvisatory practice.

Ginsberg's omission of each "(Click!)" from the published versions of his poems indicates a broader ambivalence about technology's influence on his process. Immediately after demonstrating how the clicks determine where his lines break, he qualifies this account. "It's not the clicks that I use" to arrange the poem on the page, he explains, "it's simply a use of pauses" (*SM*, 135). Line breaks refer not to the punctual click of the tape but to a longer pause in speaking, one whose duration the tape cannot record because it is turned off: "These lines in 'Wichita' are arranged according to their organic time-spacing as per the mind's coming up with the phrases and the mouth pronouncing them" (*SM*, 135). As we have discerned, Ginsberg does not maintain a rigorous correspondence between the two inscriptive technologies at hand, audio tape and paper. Instead, the clicks that mark the tape's compression of time enable him to reproduce on paper the "organic" timing of enunciation. Like the phenomenon Jacques Derrida would call "the becoming-space of time" four years later, Ginsberg's model of "time-spacing" marks the space of the page in accordance not with the mechanical timing of audio tape playback (itself a timing of the tape's movement through space) but with the time of the live speech act itself.[88] The punctual click of the recorder gestures toward a temporal dilation, a time of silence that the machine cannot record but that it still enables Ginsberg to render on paper.

The click also points toward "the force of paradox" that N. Katherine Hayles argues the magnetic tape carried in U.S. culture in the postwar years. "It was a mode of voice inscription at once permanent and mutable, repeating past moments exactly yet also permitting interventions," she writes. Unlike phonographs, which one could buy but not make at home, "magnetic tape allowed the consumer to be a producer as well" and made widely available these "powerful and paradoxical technoconceptual actors of repetition and mutation, presence and time delay."[89] When Ginsberg does not precisely transcribe the click of his tape recorder, instead letting it mutate into a line break that analogizes the time delay of actual speech, he responds to the flexibility of technique the tape recorder offers, its ability to keep a precise sonic record while also leaving room for revision. As we saw above in the work of Baraka, the play of improvisation unfolds in just such spaces between performance and revision.

The printed transcript of "Improvised Poetics" similarly negotiates the unsteady relation between audio tape and paper. The transcript includes marks that simultaneously set the limits of the recorded discourse and gesture toward an indefinite dilation of time. Shortly after Ginsberg discusses "(Click!)" as a formal cue, a paratextual note interrupts the interview: "[End of tape. Conversation resumes later.]" (*SM*, 137). Just as Ginsberg must either surmise or try to recall how long a pause in speech each click of his tape compresses, so does this note leave us to guess how much "later" the conversation resumes. This elision of time stems from a technical cause, the end of the tape, rather than an "organic" factor of the kind Ginsberg believes the tape can capture. The topic of conversation does not immediately change after the break, so maybe the discussion paused only briefly, long enough to mount a new tape. Or maybe those present took a long break, enjoyed a meal or other refreshment, and resumed much later. The same equipment that preserves this interview also sets a limit on our ability to know what was said and when. Other breaks appear elsewhere in the transcript, and these vary between editions of the interview.[90] Some of the bracketed notes summarize the substance of conversation, presumably in the name of concision. Even if we consulted the original tapes to learn how much conversation has been omitted, we might simply hear the "(Click!)" that marks the limit of recording and the dilation of its outside.

One bracketed note in "Improvised Poetics" gives an especially strong sense of the interplay between paper- and tape-based inscription technologies that structure this interview and, indeed, Ginsberg's improvisatory poems themselves. During a discussion of the poet's musical renditions of William Blake's poetry (which MGM Records would offer for sale on vinyl in 1970), the interviewer asks, "And what's your tune for that?" The transcript renders Ginsberg's answer as "[Ginsberg sings.]" (*SM*, 144). Alternatively, the 1980 edition of the interview gives us, "Ginsberg [sings.]"[91] Though formatted differently, both transcripts invite us to imagine an audible event captured on tape but only mentioned in print. Both transcripts also have a footnote about the published recording of Ginsberg's performances of Blake.[92] This note underscores the limitations not only of the printed transcript but also of the taped version that remains unheard. The commercially marketed record of Ginsberg singing Blake, now available online, might stand in for this moment of singing in the interview, but any two recordings of someone singing the same tune will differ in certain ways. On the published record, Peter Orlovsky sings along with Ginsberg while Jon Sholle accompanies on guitar, but Ginsberg apparently sings alone during the interview. The transcript substitutes a mute bracket for the sound of Ginsberg's voice and, with the footnote, insinuates a further substitution of one audio recording for another. The differences between Ginsberg's performances are at once marked in the text and impossible to arbitrate without exceeding the bounds of the transcript that marks the differences in the first place. In both Ginsberg's poetry and transcripts of his conversations, alphabetic writing at once makes visible the event of extemporaneous speech and marks the limits of any inscriptive technology that seeks to capture such an event, whether on paper or on tape.

Ginsberg's way of using the tape recorder does not accord easily with his commitment to improvisation. He takes interest in spontaneity most directly from Jack Kerouac, whose 1953 essay "The Essentials of Spontaneous Prose" gives practical instructions for improvised writing.[93] In a 1965 interview, Ginsberg calls Kerouac "the best poet in the United States" because "he's the most free and the most spontaneous" (*SM*, 50). For Ginsberg, improvisation enables a talk-oriented poetics that can be traced back to Wordsworth, who believed poetry should eschew formality in favor of "a selection of the language really spoken by men."[94] Whether either poet manages to write in a true vernacular remains open to debate, but Ginsberg praises Kerouac in similar terms of authenticity. "Kerouac told me," he reports, "that in the future literature would consist of what people actually wrote rather than what they tried to deceive people into thinking they wrote, when they revised it later on" (*SM*, 51). Given the importance of spacing and repetition for any conventional writing system, I have difficulty understanding Kerouac's paranoid prediction. If revision were truly ruled out, writing would almost cease to be writing. The tape recorder clarifies somewhat, if we understand it as an inscriptive device, a writing technology that captures language immediately as it is spoken, without the delay of composition on paper.

Ginsberg perceives spiritual benefits in the spontaneity that tape recorders apparently support, but this function differs from the mystical powers others attribute to audio tape. Burroughs and Brion Gysin, for example, use tape in their experiments with the "cut-up" method. Burroughs claims that "some of the cut-ups . . . seem to refer to future events."[95] The cut-up technique, however, involves chance operations rather than improvisation—the selection of words from a hat, for instance, rather than the spontaneous decisions of the artist—so it shares more with the mystical attachments to chance discussed in chapter 2. In Ginsberg's hands, the tape recorder enables a method of spontaneous composition that he considers freer and more authentic than writing done on paper.

Nonetheless, Ginsberg also expresses misgivings about electronics, including tape recorders. Per above, he views electronics as instruments of control and worries that they stifle the same spontaneous expression the tape recorder seems to capture. His defense of free speech on *Firing Line* is framed politically, but it concerns the media through which he speaks. In an earlier interview, Ginsberg decries a "highly centralized network of artificial communications" that leaves us "surrounded by machines and electricity and wires," making it "increasingly necessary to have a breakthrough of more direct, satisfactory contact that is necessary to the organism" (*SM*, 58). Ginsberg's desired kind of sociality requires "direct" contact between organisms, without the prosthesis of "electricity and wires." Juxtaposed in this way with various technological constraints, this organismic personhood might also provide the basis of spontaneous action the poet so values. Elsewhere, however, he describes the mind through electromechanical tropes that privilege inscription and leave little space for spontaneity. Discussing Blake and other poets, he claims that "any experience we have is recorded in the brain and goes through neural patterns and whatnot: so I suppose brain recordings are done by means of shifting around of little electrons—so there is actually an electrochemical effect caused by art" (*SM*, 32). Whatever their scientific merit, these speculations leave little room for improvisation. Ginsberg describes the brain as a causally determined, "electrochemical" mechanism that "records" stimuli, presumably storing them for later playback. This model fits neatly with the inscriptive way around improvisation described earlier. Even if the neural mechanism somehow produces free agency capable of spontaneous action, the brain that witnesses improvisation subsumes it within a logic of recording, of inscription, that stifles its spontaneity. If I witness an improvised performance that nobody records, can I still regard it as improvised every time I remember it? If so, the experience of organic memory differs from electronic recording in some way that Ginsberg's theory ignores.

Ginsberg also deploys mechanistic tropes for poetic composition, and these make it difficult to say how his own compositional equipment leaves room for improvisation. He complains that poets writing in accentual meter "just fill in a form which is made up in advance, like a bureaucratic form" (*SM*, 107). Belgrad links such complaints with the Beat writers' interest in improvisation. He argues that Beat poets "oppose the culture of corporate liberalism with a spontaneous prosody."[96] Ginsberg therefore uses technological language to describe poetry that lacks spontaneity: "If the accent becomes automatical mechanical the rhythm becomes automatical mechanical, and the emotions therefore are automatical and mechanical and the intellect is automatical mechanical" (*SM*, 107). The assonance and stilted rhythm of this phrasing dramatize the mechanistic effects Ginsberg derides. He views accentual meter as a processing machine that regularizes language, making it "automatical" rather than free and spontaneous. This objection comports with his broader opposition to technocratic liberalism, military industrialism, and social alienation, but the technophobic rhetoric opens difficult questions about his use of audiotape. If he dislikes "mechanical" ways of processing language and worries that "electricity and wires" prevent "satisfactory contact" with others, then why does he embrace the reels and sprockets of the tape recorder as equipment for poetry?

Ginsberg expresses similar ambivalence in his most direct comments about compositional equipment. On one hand, he sometimes suggests that technology strongly determines poetic forms: "The page determines the length of the line. It's an arbitrary thing" (*SM*, 125). Hence, narrower pages produce shorter lines and vice versa. Any technology that constrains form this directly would seem a poor platform for poetic improvisation. On the other hand, Ginsberg elsewhere downplays the influence of technology, as when he claims that the "(Click!)" of the tape does not directly occasion a line break but refers to the indefinite interval between recorded phrases. When an interviewer suggests that the tape recorder opens "a direction away from the printed poem," Ginsberg insists that "the tape recorder's not much different from writing in a notebook" (*SM*, 157).

Both paper and audiotape use inscription to record the poetic invention. Tape may capture live speech more directly than paper does, but it remains an inscriptive mechanism against which improvisation develops its meaning. As Michael Jarrett argues, "we understand composition as 'not-improvisation,'" so improvisation "is ultimately a consequence of writing; it is thinkable only within textuality."[97] We encountered a similar perspective in Moten's dialectical theory, which positions improvisation as a kind of limit concept, discernible only in tension with its others. Ginsberg's tape recorder, by this account, textualizes his improvised speech, making it a basis for composition. Ginsberg's own statements about improvisation and technology espouse a less nuanced view of the former, but his ambivalence about the latter suggests he does sense that electronics simultaneously enable and constrain poetic improvisation.

Ginsberg believes inscriptive devices record not only the products of improvisation but also, at a formal level, the technique's founding moment of uncertainty, of not knowing what comes next. Just after claiming that page width determines line length, he argues that "what also determines where the line breaks is: when the thought breaks."[98] Ginsberg does not mean that each line contains an intact unit of thought, as one might suppose, but that lines break where thought *breaks down.* Discussing a line that ends with "total," he recalls: "I didn't know what was going to come after 'total,' total, total what?" (*SM,* 125). He therefore breaks the line and awaits the next phrase. Even when composed on paper, the poem's form bears traces of the momentary uncertainty at the heart of improvisation. What structures the poem is not some rational plan nor the equipment used to make it, but the open, unforeseen present. The technologies of poetic inscription may provide an "absolutely logical scientific notation of event," but this event of poetic creation remains spontaneous in itself (*SM,* 154). Elsewhere, Ginsberg does allow that audiotape has advantages for improvisation: "The tape process got me talking more—well, I was able to get more fugitive things going" (*SM,* 137). This concession to the automatism of tape matches our intuitive sense that it captures enunciation on the fly, preserving utterances that might not survive the delay of pen- or keystroke. Ginsberg's mention of "fugitive things" might seem to anticipate Mackey's idea of fugitivity, a term that evokes both the fleetingness of improvisation and the history of African American enslavement, with its dialectic of flight and capture. If Ginsberg's improvisations refer us to a history, though, that history is technological.

The aim of preserving fugitive language motivated even the earliest inventors of sound recording. One such inventor, Édouard-Léon Scott de Martinville, patented a device called the phonautograph, which traces sound waves onto a rotating drum coated with lampblack. The recordings could not be replayed, but they enabled visual study of sound waves.[99] Even in 1857, Scott understood the fundamental link between sound recording and the fleeting event of improvisation. Speculating on the future of recording technology, Scott asks,

> Will the improvisation of the writer, when it emerges in the middle of the night, be able to be recovered the next day with its freedom, this complete independence from the pen, an instrument so slow to represent a thought always cooled in its struggle with written expression? I believe so. The principle is found.[100]

The movement of the pen delays and interrupts the flow of thought in language. By recording speech in the time of its utterance, Scott believes, successors of the phonautograph can preserve improvised language directly. Commenting on this passage, Jonathan Sterne claims Scott's theoretical fixations kept him from devising means of audible playback, as Edison and others soon would: "His inability to see even the Edison phonograph as a major improvement on his own device was an artifact of a monomaniacal focus on writing, on the *product* of the machine, overlooking its more processural dimensions."[101] A history of sound reproduction such as Sterne's must certainly address processes other than inscription, including the tympanic registration of airborne sound, which Sterne identifies as Scott's innovation. But from Scott's day to ours, various kinds of inscription have provided the enabling supplement for sound recording and replay. As Scott's mention of improvisation confirms, the supplementary work of inscription proves especially important for theorizing spontaneity in language. For this reason, when Ginsberg avows that "written language is a mind crutch," we can deduce that he views tape-recorded language in the same way, as an inscription of language that aids (and thereby weakens) the memory (*SM*, 156). He recognizes a fundamental opposition between improvisation and the inscriptions that capture it, but he also sees the latter as indispensable means to preserve and disseminate his spontaneous poetry.

This view of inscription as a "crutch" for improvisatory performance may resolve questions about poetic production, but it does little to explain the reception of improvised poetry. In this area, a fundamental question remains: if an improvisation is recorded, whether on paper or tape, does it still count as improvisation when it is later read or replayed? If so, is its improvisational character something stylistic, aesthetic, rhetorical, ontological, or what? When we listen to recorded performances of poetry, it can seem that audio recordings capture the live moment of improvisation in a way that alphabetic writing cannot. Perhaps, as my comments on Baraka's paratexts suggest, a poet's improvisations are most evident in his extemporaneous departures from the written text he reads, in his elisions, stammers, and impromptu revisions. But Ginsberg's account of his performance technique confounds this claim:

> When I get up and read those units I just make believe I'm trying to think of the next phrase. And then I come out with the next phrase. . . . So it leaves the mind of the reader or listener to be hovering with mine. (*SM*, 126–27)

On one hand, audiotape does capture Ginsberg's feigned indecision. On the other, the poet admits his hesitations are "make believe," false improvisation, when he has in fact carefully prepared. He even plans to pretend his speech is unplanned, a common and powerful technique of stage performers, but not really improvisation. The recording, then, may fail to capture improvisation not only because it sounds the same every time it is replayed but also because, in the time of performance, Ginsberg pretends to improvise while actually adhering to the script of the poem.

This predominance of inscription helps to explain the effects of Ginsberg's habit, starting in the late 1960s, of singing and chanting while he performs. His various songs and chants have important spiritual motives (benediction, pacification, purification), but these more musical practices also help Ginsberg to emphasize the liveness of his performance.[102] He may improvise his tape-recorder poems at the moment he speaks into the microphone, but thereafter, his language gets subsumed in an interplay of inscriptive technologies that dampen its spontaneity. In song and chant, by contrast, the voice moves away from language and approaches the condition of bare sound.[103] The song or chant enables Ginsberg, at the time of performance, to improvise beyond his prewritten poems. His use of chanting as a paratextual element of performance is evident in his 1968 interview on Buckley's *Firing Line*. Buckley deescalates a heated moment of debate by suggesting that Ginsberg read another poem, and the poet responds by asking, "Why don't I sing instead?" He then produces a small harmonium and performs the Hare Krishna mantra for approximately one minute. Although Ginsberg does not deviate from the sixteen-word script of the chant, he improvises through melisma and altering timbre. He prefaces the chant by dedicating it to "the preservation of the universe instead of its destruction."[104] Later he and Buckley debate the suitability of chanting as an alternative to war in Vietnam. They seem to agree that both poetry and chant are distinct from the more sober points of political argument and that, in its broader sense, "song" offers an alternative to the violent discourse of the state. As a means of deviating from the script, chant helps Ginsberg to maintain this poetic difference from rational disputation.

Ginsberg's chanting represented a significant departure from the way he performed at the time he gained fame, and this change to a more spontaneous performance style persisted for the rest of his career. Those who discuss Ginsberg's early readings often mention their energetic intensity. Raphael Allison, for example, hears an "emphatic, oracular power" in Ginsberg's 1959 performance of "Howl" in Chicago. The Whitmanian structure of this and other early poems, with their long lines and ceaseless anaphora, lends Ginsberg's voice a mounting verve in such early readings. Still, these are readings: the poet perceptibly recites a script and gives little evidence that he aims for spontaneity, as he later would. In Allison's account, the powerful reading of "Howl" in 1959 takes the "measure of the poem's underlying movements of feeling, which build over three parts in a slow crescendo." In other words, Ginsberg plans and modulates his performance according to a preexisting script. Allison notes that, by contrast, his 1971 performance of "Howl" in San Francisco is "lively, comic, edited (possibly spontaneously), interspersed with comments" about the poem—imbued, in other words, with the kinds of textual instabilities and paratextual noises that mark a performance as improvisatory.[105]

Nearly a quarter century after singing Hare Krishna on Buckley's program, Ginsberg remained dedicated to chanting. He began a 1992 reading at the Naropa Institute by chanting the syllable "ah" at various pitches for several minutes, again accompanied by the harmonium. Before starting, he describes the chant as "symbolic of purification of speech, like a white sound in the throat that eliminates all conceptualization."[106] The chant thus purifies the mind of conceptual clutter, a release from overly rational thinking. Ginsberg is far from alone in taking the voice's potentially nonverbal musicality as an outside of rational discourse, where poetry finds its limit condition as pure sound. In the context of studying improvisation, it seems especially apt that he calls the chant "white sound," a synonym for white noise, which masks disruptive sounds. In an important sense, as Ginsberg knows, his chants are not white at all: he appropriates them from Indian and other non-Western cultures. Insomuch as Ginsberg's performances assimilate these nonwestern spiritual practices of spontaneity, his chanting may also function as white noise in the sense that it drowns out the nonwhite cultural histories from which he draws. As the previous section made clear, Baraka and many others perceive the political urgency of discussing improvisation as a practice that might be called "black sound," a sonic tradition of nonwhite cultures from which the most influential practices of improvisation emerge.

Ginsberg celebrates poetic improvisation for its difference from rational planning and conceptual predication. In Ginsberg's printed poems and recordings of his performances, the discourse of improvisation fundamentally relies upon inscriptive technologies that seem to challenge the very idea of unplanned action. The devices that capture and preserve poetry also threaten to stifle its spontaneous verve, making it repeatable and predictable. Yet poetic improvisation constantly evades determination by its technological conditions. As Jacques Coursil argues, "in improvisation, one has to invent an event . . . create the advent of a present" that is unforeseen, not predetermined.[107] The spontaneous poetic event outstrips the technologies that help to determine what theories of improvisation and ways of reading it seem available. Improvisation thus sets up a familiar relation between poetry and its equipment. In the hands of poets, devices that seem rigorously informational and rationally organized sustain less orderly ways of thinking. Seen in their technological contexts, the resulting poems suggest that electronics may not have strictly rationalizing effects after all. This ambivalence about the rational vocation of technology shapes the production and dissemination of Ginsberg's poetry. He talks spontaneously into his tape recorder, but the machine organizes his speech into a determinate pattern on tape, making it predictable. Even as he argues that tapes and notebooks are not so different, he admits that tape recorders help him capture more "fugitive," more spontaneous language. Likewise, when we hear Ginsberg performing for an audience, the minutest audible quirk of his reading can sound the same every time we listen, so even a wildly spontaneous moment eventually becomes rote. Yet during most of his career, Ginsberg supplemented the recital of prewritten poems with paratextual songs and chants that open the voice to spontaneous musicality, much as Baraka's do.

The online archives of recorded poetry give further evidence that audio recording does not wholly rationalize spontaneous poetry. The archives are a mess, so full of incomplete, spotty, and inaccurately catalogued material that they cause confusion, frustration, and surprise more often than they give a sense of stringent order. To take one example, one of the largest online collections of Ginsberg recordings is the Naropa Poetics Audio Archives, the product of a discontinued collaboration between Naropa University and the Internet Archive.[108] This site contains an item titled "Allen Ginsberg and Philip Whalen reading, June 1975."[109] The notes on this item list its runtime as 48:20, but the audio file itself is only 3:30 long. It includes only a sound check and the start of Anne Waldman's introductory remarks. A fuller recording of the event, 1:36:41 long, is available on the separate Naropa University Archives website.[110] Even this recording, however, omits the final portion of the performance, during which Ginsberg played copyrighted material. The Naropa archivists have retained the original tape of the full performance but do not make this final portion available online for fear of violating copyright, the same reason that Naropa ended its collaboration with the famously prosharing Internet Archive, whose platform made it too easy to download copyrighted materials.[111] Stranger still, the longer recording on the Naropa University Archives site omits the sound check and begins with Waldman's introductions. As a result, the first 2:20 of that fragmentary, truncated recording in the Internet Archive can be heard nowhere else online, only there and on the tape in Naropa's local archive. That shorter recording begins with Ginsberg playing the harmonium and singing:

> Can you hear me? With the words, or does the harmonium sound come too loud? I don't want to get too close to the microphone, or should I get my head back [indiscernible][112]

Ginsberg spends the next moments asking the audience how close his harmonium should be to the microphone so that it does not drown out his voice. His song about the microphone may sound the same every time, but he clearly improvised it on the spot, a fact underscored by its imperfect capture on tape. The harmonium does indeed drown out some of the words he sings, just as he fears. Further, this short audio file stands as a mislabeled fragment in the digital archive; it refers us to a larger, absent recording that, even when we find it in another digital archive, remains partial, cut off at both ends, and excludes the paraperformative song about a microphone heard here. As the sound-check song suggests, Ginsberg's improvisations open challenging questions about the interfaces among the voice and the technologies that capture and preserve it—among the digital archive, the microphone, and the spontaneous song that is always in some sense about the microphone, about its own conditions of recording. By staying receptive to the surprises and disorder of poetry archives emerging online, it may be possible to theorize these interfaces in ways that do not always subordinate spontaneous action to the equipment that preserves it.

THE IMPROVISED ARCHIVE

The online audio archives to which I refer in this chapter give a sense of improvisation I have not yet addressed at much length, the sense of provisionality. When we call certain things improvised, such as shelters, mechanical repairs, or explosive devices, we refer to their makeshift, temporary, suboptimal, or amateurish character. This dimension of the chapter's key term, rather than its relation to spontaneity, comes to the fore in today's online archives of literary audio. On one hand, these archives fill a certain need, provide a substantial convenience, and serve their purpose passingly well. On the other hand, the largest and most popular online archives of literary audio seem improvised in significant ways, constructed and maintained ad hoc, rather than systematically planned and organized. To call an archive "improvised" in this sense may sound like derogation, but that is not my primary aim. Although today's online archives certainly have shortcomings, these imperfections lend them a more flexible, casual, provisional air that many prefer over the orderliness and stability of conventional archives. I will close this chapter with a brief discussion of PennSound, the best known online archive of literary audio, in order to describe the advantages and limitations of such improvised collections.

Before launching PennSound with Al Filreis on January 1, 2005, Charles Bernstein wrote a "PennSound Manifesto," dated 2003, to guide the archive's design and development.[113] By measuring the six principles of this manifesto against today's fully developed archive, it becomes possible to discern how the improvised aspects of the site, its unplanned elements and its deviations from the stated plan, influence its function. To summarize,

Bernstein proclaimed that files on PennSound should: (1) be free and downloadable to a user's local hard drive; (2) utilize "MP3 or better" file formats; (3) "be singles," separate files for individual poems, rather than whole performances; (4) follow a uniform system for filenames that would include the poet's name, poem title, location, date, and so on; (5) "embed bibliographic information in the file" using ID3 tags; and (6) be indexed so that search engines and library catalogs can access the archive.[114] Some of these principles are quite constructive, but others are oddly conceived. PennSound does an excellent job of meeting the first and third criteria, both of which make the archive more useful. The entire site remains free and public. All audio files can be streamed from the server or downloaded, although it is not as easy to download the videos that the archive has recently begun including. Almost every reading is divided into "singles," giving the option to download audio of individual poems or of the whole performance. At this basic level of usability, PennSound meets its goals quite well.

The usefulness of other precepts in Bernstein's manifesto is less clear. For example, he calls for "MP3 or better" audio files in order to maintain "reasonably high quality sound," but MP3 uses lossy data compression, making it less suitable for high-fidelity audio.[115] Bernstein does not suggest a minimum or optimum bitrate, which determines the fidelity of a given MP3. He prefers MP3 over RealAudio because the latter "is a proprietary format," but MP3 is also a patented format.[116] If PennSound wants open file formats it might use Ogg Vorbis, FLAC, or WAV, formats that Wikipedia and other public resources prefer because of MP3's patent encumbrances. Meanwhile, PennSound contains some old recordings of significant interest but whose fidelity is inevitably low. The recordings of Robert Frost, for example, were first made on aluminum platters on May 5, 1933, later dubbed to reel-to-reel audiotape, and finally digitized for PennSound.[117] One often wishes for higher-fidelity audio in PennSound, especially in recordings made before the 1990s, but in most cases, the shortcomings lie with the source media, microphones, or other recording equipment, not with the destination file type.

Likewise, Bernstein's fourth and fifth principles, about filenames and ID3 tags, set reasonable but not always useful guidelines. The filenames at PennSound generally follow the format Bernstein proposes in his manifesto: lastname-firstname_cut-number_title_place/series_date.mp3. Some filenames deviate from this formula, however. Turning again to the Robert

Frost holdings, his author page states that certain recordings were made on May 5, 1933, but the "date" position in the filename lists only the year.[118] Other filenames provide a full date, while some include no date at all, even when the date is stated on the site. Last but not least, the ID3 tags (also known as metadata) for files downloaded from PennSound contain information about each file, but sometimes the details in the ID3 tags remain vague or incomplete. The recordings of Charles Olson, for example, include something called "Christmas Tape Reading, 1957." Although the phrase "Christmas Tape" is repeated in the relevant filenames and ID3 tags, no explanation is given. Should we suppose the reading happened around Christmas, 1957? It seems of very little use to know the name of some series or event to which this recording corresponds if we do not also know the date or place of the reading. Hence, some of the principles intended to organize the PennSound collection and ensure the high fidelity of its holdings do not always prove equal to that task. These imperfections do not diminish the substantial value of the PennSound archive, but they do texture the experience of using it.

Surprisingly, Bernstein's manifesto does not address some factors that have proved key to the structure of the PennSound project. For example, he does not explain why the archive's organizational scheme emphasizes the names of individual writers. The PennSound homepage offers a navigation menu in the left sidebar, and the first option there is "Authors," which links to an alphabetical listing of authors in the archive. The navigational menu does offer other ways to explore, including by "Series," which lists recurrent reading series, or "Classics," in which moderns read works by Chaucer, Sappho, Keats, and others. Where an author's name appears in these other listings, the name often links to her individual author page, but the author pages do not link to every "Series" or "Collection" page where the author appears. The individual author pages thus provide the grounding organizational logic for the archive. There is no simple way to sort recordings geographically or chronologically, for instance, if you want to hear readings from a given city or year. As a result, the archive's search function becomes more important, since it can help to find recordings by place or date. The embedded search tool, however, is simply a domain-specific Google search; it lacks advanced search options, as well as a means to sort or filter results. The archive partially meets the sixth criterion of Bernstein's manifesto, which says PennSound must be "indexed" and "retrievable both from a library catalog under an author's name and via web search engines."[119] Although PennSound author pages do appear in Google searches of the open web, the archive does not seem to be included in any library catalog search, including the libraries of the University of Pennsylvania, PennSound's home institution.

The emphasis upon individual authors at PennSound does not comport with its founders' views on authorship. Both Bernstein and Filreis have long supported experiments to disrupt, escape, or decenter the sense of a personal, expressive voice in poetry. Indeed, if PennSound exhibits a preference for one kind of writer over another, it prefers avant-garde poets of the antiexpressive sort, including Language writers, conceptualists, and the like. In this digital archive of authors' voices, the reemergence of authorial individualism at a structural, organizational level, despite an almost doctrinal aversion to such conventions on the part of its creators, gives a sense of the power with which expressivist tropes of the lyric reassert themselves through new poetic technologies such as the online archive.

Within each PennSound author page, recordings are arranged chronologically, but with a casualness and inconsistency that, as one uses the archive more, increasingly appears to be its dominant mood. Recordings on Ginsberg's page appear in chronological order, with readings from the late 1950s at the top and those from the mid-1990s at the bottom, followed by undated recordings. Baraka's page, on the other hand, is organized in reverse chronological order, with readings from the 2000s at the top and those from the mid-1960s at the bottom, again followed by undated material. Roughly half the author pages are organized chronologically, half the reverse. The recordings on other pages, including those for Susan Howe, Kathy Acker, Myung Mi Kim, and Chris Funkhouser, are arranged either randomly or chronologically but with errors. These inconsistencies prove easy enough to negotiate, whether one is browsing the archive casually or looking for something specific. As PennSound continues to grow, however, and as it remains the most notable source of literary audio online, the effects of such organizational errors and inconsistencies will likely become more pronounced, making it increasingly difficult for human users and software to interface productively with the archive. These imperfections encourage users to interact with the archive in a more casual, unsystematic way. The visitor who simply wants to poke around and discover something new will have no trouble, even if these organizational issues make PennSound less suited for more formal, scholarly uses.

In a 2006 article about the recently founded PennSound archive, Bernstein identifies one goal of the project as delivering performed poetry in more casual listening environments. With an online archive of the kind he imagines, "the experience of listening to poetry would be far more mobile and portable than it has been." He envisions users listening to poetry "in the car or at the health club or on the plane or walking around the city or sitting at the beach."[120] Allison is correct that "sites like PennSound . . . [are] shifting the bank of primary materials students and scholars can study away from solely print-based areas," but as the site's improvised structure makes clear, its founders' intentions are not so academic.[121] Written in the mid-2000s, Bernstein's fantasy of mobile, casual, ubiquitous listening can be read as a response to the consumerist fervor over the Apple iPod, released in 2001, and the MP3 mania it fueled. Bernstein's effort to bring recorded poetry into daily life is also, however, a productive reaction against the dusty archives of obsolete audiotape formats to which such recordings had been mostly confined. Indeed, the online archive makes recorded poetry immeasurably easier and more common to hear. In this respect, PennSound's founders have rendered an enormous service to Anglophone poetry and have substantially reshaped its textual conditions.

Bernstein and Filreis's aim to casualize the experience of hearing poetry chimes with their broader opposition to authority and formality in the literary world. Bernstein coined the phrase "official verse culture," a term of derogation now familiar among contemporary experimental poets who oppose what they consider an entrenched mode of expressive lyricism.[122] Filreis, meanwhile, regularly teaches a massive open online course, "ModPo," which has introduced many thousands to modern and contemporary poetry, free of charge. PennSound itself now includes, alongside the archive, an increasingly active selection of blogs and news feeds that make the site feel more like a kind of online magazine for audio poetry and poetics, and less like a slowly growing archive. These democratizing, casualizing impulses certainly deserve praise, but they become increasingly ironic. Bernstein and Filreis may once have qualified as scrappy opponents of the elitist literary establishment, but now they hold named professorships at an Ivy League university. If the Donald T. Regan Professor of English and Comparative Literature and the Kelly Family Professor of English at the University of Pennsylvania do not qualify as members of "official verse culture," one wonders who does. By the same token, PennSound does make it possible to hear poetry outside the confines of live readings and library archives, but it has also become the best known and one of the largest online archives of literary audio, making it both a popular destination for poetry aficionados and an important research tool. Vital resource though it may be, the improvised character of this archive—its organizational inconsistencies, limited searchability, and integration with a relatively less stable poetry-news infrastructure—diminishes its usefulness for research. But given its founders' ambivalence about formality in the world of poetry, maybe that is the point. If the readings of poetry and its equipment in *The Poem Electric* have demonstrated anything, they have shown that poetry does not always serve the ends of information and rationalism. Perhaps poetry archives like PennSound, if they serve the ends of poetry, need not remain entirely orderly themselves. Instead, if such archives appear improvised, thrown together on the fly, then perhaps they equip us all the better to hear the traces of poetic improvisation they preserve. The *longue durée* of the conventional archive, its commitment to stability and endurance, seems likelier to dampen the lively events of improvisation. By virtue of their imperfections, PennSound and other online poetry collections invite users to see them as places to get lost in, to discover something unexpected, or simply to linger, instead of engaging with poetry in a more organized, planned way. In the opinion of this scholar, at least, the world would do well to engage with poetry in less formal, more spontaneous ways.

Conclusion

Lyric and Objecthood

This book has traced the persistence of lyricism in the era of poetry's electrification. Although electronics might seem to submit language to a highly rational order, *The Poem Electric* shows how these technologies sustain lyric poetry's exemption from rationalism, its basis in aesthetic modes whose value lies in their difference from knowledge. Such exemption from rational thought proves quite flexible, however, and ultimately supports alternative ways of knowing: affective response, for instance, enjoys immunity from skeptical interrogation but nonetheless underwrites claims of certainty, and improvisation entails *not* knowing what one will do next but also requires a certain know-how. Beyond these insights about poetry's epistemic conditions in electronic environments, the readings in this book more generally indicate that electronics do not cleanse poetry of its lyrical intensities—of its expressiveness, playfulness, or intimacy, for instance—so thoroughly as one might presume. Instead, poets and literary critics have made inventive use of electronics to clarify and rework the familiar parameters of lyricism. In this way, the postwar and contemporary poetic avant-gardes share more with the lyrical tradition than they may wish to acknowledge, and the instruments of their apparent deviance from conventions have in fact supported a highly productive rearticulation of familiar lyric modes.

To give this account, *The Poem Electric* has developed interpretive methods that hew closely to the material dimensions of a poetic text without abandoning more conventional hermeneutics. The difficulty of striking such a balance has become a familiar motif of lyric reading: the objecthood of a poem becomes most directly apprehensible at the moment

when it appears as dumb matter, signifying nothing, and its verbal significance leaps forth most powerfully just when it exceeds the material particularity of its technological condition. A lyrical pathos emerges from the
interplay between these two aspects of a poetic text, material and semantic.
This interplay attunes readers to what I call lyric objecthood, the poem's
status as both a material thing and a verbal expression. By remaining attentive to this dual condition, I have sought to produce detailed technological
histories of poetry without framing lyricism as a merely secondary or accidental effect. For many who would read literature in terms of its technological dimensions, it remains difficult to avoid determinist frameworks that
subordinate literature to technology and unduly privilege historicist reading
strategies. I have therefore found it useful to consider this project a rhetorical poetics, rather than a history, because rhetorical interpretation accom-
modates both technological and more properly literary ways of reading.

This conclusion draws out the implications of these reading strategies
for some recent poems that rely conspicuously upon computers, works
that qualify decisively as e-literature or digital poetry. For two reasons, I
have said relatively little about this kind of poetry so far. First, others have
already published some fascinating research about technologically innovative poetry, but less has been written about the subtler, more pervasive
effects of electronics upon poems that, like many of the poems in this
book, do not seem glaringly "digital." Second, a lot of recent e-literature,
including some that has garnered significant critical attention, frankly does
not strike me as very successful literature, e- or not. Sometimes the technical inventiveness of such work distracts from its literary mediocrity. We
remain in the very early days of computer-based literature, however, and
we are sure to see more successful efforts in the future.

The following pages therefore discuss three computer poems whose
shared aesthetic traits do reward close attention. The first two are *Dakota*
(2002), by the duo known as Young-hae Chang Heavy Industries (YHCHI),
and *Project for Tachistoscope {Bottomless Pit}* (2005), by William Poundstone. These have already occasioned some critical commentary, which I
will link with my own claims about the afterlives of lyricism.[1] The third
piece, Tan Lin's *Mastering the Art of French Cooking and Systems Theory*
(2015), is stylistically similar to the other two but has not yet gained as
much attention.[2] All three poems work by flashing single words or short
phrases onscreen in rapid succession. The words appear in plain, bold
type, most often black. The texts of the YHCHI and Poundstone pieces

are original, written by their authors. *Dakota,* which its authors say is an adaptation of Ezra Pound's *Cantos* I and II, tells a disjunctive tale about a debauched road trip in the American West. *Project* tells a strange, looping story about a bottomless pit. Lin's piece, by contrast, juxtaposes words from two found texts, Julia Child's famous cookbook *Mastering the Art of French Cooking* (1961) and a 1991 lecture, "System as Difference," by the sociologist Niklas Luhmann. The two earlier pieces also have musical accompaniment: *Dakota* includes an Art Blakey tune whose rhythm the flashing words match, while *Project* plays a low, throbbing drone. The Lin piece is silent. These three poems share two important structural traits: first, they flash words onscreen so quickly that viewers cannot catch every word, and second, they provide no controls by which the viewer might "navigate" through the piece or control playback, for instance by pausing or rewinding. Through these shared traits, I will argue, these poems inventively transpose lyric conventions into computer environments, thereby exploring how electronics alter a poem's objecthood. My conceptual guide in reading these poems is Michael Fried's essay "Art and Objecthood" (1967). Although Fried interprets minimalist painting and sculpture rather than poetry, his commentary anticipates the difficulties of describing the lyric objecthood of computer poems.[3]

One such difficulty has to do with the complexity of computers. To address the objecthood of a poem on paper, or even a performance captured on audio tape, is much simpler than to describe the material substance of a PDF on my computer. Writing this sentence with a pen, I can easily remark how the nib deposits ink as it moves across the page. An equivalently materialist description of typing this sentence on a laptop would be significantly more arduous. This difference makes it challenging for electronic poetry to figure its objecthood. As Fried observes, artworks that most effectively call attention to their objecthood tend to "assert the values of wholeness, singleness, and indivisibility," a style of simplicity that underscores the basic materials of the work.[4] This is why critics associate the objecthood of Dickinson's poems with her simplest mark, the dash, rather than a more elaborate pen stroke. In important ways, however, the material structures subtending e-literature lack this simplicity, making it difficult to remain in touch with the objecthood of electronic texts. Indeed, scholars in decades past who breathlessly construed "the virtual" as a kind of computational immateriality may have done so in part as a defensive response to a technology so complex it seems indescribable. Of course, paper itself is far from simple, and historians of literary technologies have begun to tell its story.[5] But the old vocabulary of lyric objecthood has equipped us with familiar ways to describe the conjunction of verbal expression with the physical deposition of marks on a page. The objecthood of a poem on paper enjoys a self-evidence lacking in even the simplest computer processes. As Clement Greenberg puts it in a passage Fried cites, "the look of machinery . . . does not go far enough towards the look of non-art, which is presumably an 'inert' look," to invoke the simple objecthood of the artwork (152). If Greenberg is correct, then writers who invoke the objecthood of electronic texts face the challenge of doing so without rendering the complexity of the computer so faithfully as to distract from its substance. The annals of e-literature may still await their equivalent of the Dickinsonian dash, a mark so elemental that it refers at once to the substance of a literary medium and to the passions of its poet.

Some of the best computer poems, including the animations by YHCHI and Tan Lin, pursue just such a style of visual simplicity. Instead of reveling in the computer's ability to produce complex visual effects, these poems display words as plainly as possible, rendering them in bold, black letters against contrasting backgrounds, usually white. They do employ simple animations to move through the text, but they still make relatively little spectacle of the visible word. Their typefaces do not look three-dimensional, for instance, nor do they simulate the look of paper by means of artificial textures or other skeuomorphs. They offer few visual embellishments or stylistic diversions. Instead, the animations and other design choices emphasize the simple, direct appearance of words onscreen. Lin's piece uses two common typefaces, Courier and Helvetica. *Dakota,* meanwhile, uses a modified version of Monaco, an equally simple, sans-serif type, using only the capital letters of this typeface, and it substitutes the "zero" sign, which features a diagonal crossbar, in place of the capital "O." This figure, with its techy diagonal bar, invokes the zeros of the binary code that underlies *Dakota* itself.[6] The zero thereby gestures toward another logic of simplification: anything that appears on a computer can be reduced to ones and zeros, as it were, to the simple binary difference at the heart of modern computation. In multiple ways, then, these computer poems embrace a visual simplicity that invokes their objecthood, their material basis, more powerfully than elaborate visual effects would. To display even a single word on a computer does involve an astoundingly complex series of physical processes, but the bracing immediacy of a bold, black word on a white background calls to mind the materiality that subtends any verbal process. The style of verbal simplicity brings these computer poems toward a condition of generic exhaustion that Fried associates with the artwork's objecthood. He views painting as "an art on the verge of exhaustion" because it has confronted the limiting conditions of its genre, such as the flat, rectangular canvas (149). To paint on different shapes, Fried warns, would "merely prolong the agony," for the disruptive question about generic conditions has already been asked. Such questions lead to situations of generic exhaustion in which artists adopt the formal precepts of other genres. For instance, by working on nonplanar or nonrectangular surfaces, the painter acts more like a sculptor. Likewise, the emphasis upon the shapes of words in some concrete poetry and in the computer poems I have described lends them an iconic quality, subordinating their semantic sense to their abstract visual forms so powerfully that their verbal function begins to wane. In the process, these works exhaust a generic condition of poetry: the requirement that it consist of words, that it signify verbally. Hence, the exhaustion of a genre entails both a breach of its boundaries and a clarification of its fundamental precepts.

The Poem Electric has described just such generic exhaustion in what I call the afterlives of the lyric, wherein a lineage of avant-garde writers clarify and reassert the core precepts of lyricism precisely by exceeding them. These practices of postlyrical exhaustion continue in some computer poetry. Lin's piece, for instance, draws all of its words from two nonfiction texts, neither particularly lyrical, but by setting up randomized juxtapositions of words from these texts, he enacts a chance-based lyricism akin to Jackson Mac Low's. Likewise, *Dakota*'s jazz soundtrack blurs the lines between music, poetry, and sound art. But here too, the music provides a rhythmic emphasis for the words, making the work more expressive and affectively intense. Fried's essay makes clear how easily we can misunderstand such generic hybridity. Although he praises minimalist sculpture for reflecting upon its own objecthood, he criticizes the same work for staging the encounter with sculpture as a kind of theater, a performance or presentation. Fried does not view such genre-bending work as productively breaking barriers, empowering the artist with techniques previously unavailable in her genre. Rather, he sees it as a sign of exhaustion that clarifies the central values of an artistic genre. Generic crossings of this kind show that "the individual arts have never been more explicitly concerned with the conventions that constitute their respective essences" (164). A book like *The Poem Electric* could easily have claimed that its poets free themselves from the constraints of lyric norms. Indeed, such claims have been made persuasively in response to Stein, O'Hara, Mac Low, and others discussed earlier. As Fried's analogous points about sculpture and painting indicate, however, many apparent departures from lyricism in fact intensify the influence of the lyric's organizing conditions. That intensification, unfolding at the limits of the lyric genre, continues in the most successful computer poems. In particular, their treatment of temporality, personality, and gesture heightens the reader's awareness of their lyric objecthood.

Fried identifies in painting and sculpture an effect of temporal suspension also commonly associated with lyric poetry. The plastic arts, he argues, enjoy a "presentness and instantaneity" that outstrips any timeframe or processual sequence (167). We might understand this presentness in two ways: a sculpture conveys a sense not only of its physical presence, its objecthood, but also of its subsistence in "a continuous and perpetual *present*," an endless *now*. Fried claims that other arts, including music and poetry, aspire to this effect, and he specifies that "in poetry the need for presentness manifests itself in the lyric poem" (172n23). Indeed, Jonathan Culler describes "the lyric present" as "the fundamental characteristic of lyric," and he links this temporal structure with a variety of lyric norms, including brevity, iterability, and invocational rhetoric.[7] The computer poems at hand eschew such traditional techniques, but they find other ways of producing an effect of presentness. As the words flash onscreen, these poems give no opportunity for manual interaction through keystrokes or mouse clicks. There are no playback controls—no pause, rewind, fast forward, or scrub bar. In the era of YouTube, this omission is quite striking. The inability to adjust playback lends these pieces an effect of ceaseless temporal flow, of presentness. Lacking strong narrative sequences, they unfold in the time of their animation. Furthermore, all three pieces display successive words too fast for the viewer to catch them all; some words inevitably get missed. What Jessica Pressman observes of Poundstone's work, that it "explores and exposes how reading machines use speed toward poetic ends," can be said of the poems by Lin and YHCHI as well.[8] Poundstone titles and models his poem after the tachistoscope, a lab instrument designed to display images too fast for conscious visual recognition. Lin takes this strategy to an extreme by setting his software to display 1,200 words per minute, not too fast for a normal computer monitor to render

but too fast for a human eye to read. The speed of these poems underscores their objecthood. The experience of *not* seeing some of the text
figures the unimaginably high speed of the electronic processes that make
these poems visible in the first place. We encounter each poem's lyric
objecthood, in other words, at the limit of its legibility in a continuous,
rapidly unfolding present.[9]

Through this effect of presence, these computer poems arouse feelings of subjective distance and confrontation that are recognizably lyrical.
Fried argues that the size and "unitary character" of minimalist art makes
it less approachable; its imposing presence "*distances* the beholder—not
just physically but psychically" because the work exceeds the scope of
human thought and experience (154). The mere presence of such an artwork, like that of a person, can place certain demands upon us. Viewers
have the impression of "being distanced by the work" because it "demands
that the beholder take it into account, that he take it seriously" (155). Such
encounters with alterity and impenetrability yield a feeling of distance or
absence. As Fried writes, "Being distanced by such objects is not . . . entirely
unlike being distanced, or crowded, by the silent presence of another *person*" (155). His language here, especially the conflation of "being distanced, or crowded," recalls the lyrical interplay of presence and absence
in Frank O'Hara's work. In his poems, a name on the page figures the absence of the person named and, through this substitution, makes the poem
personal in O'Hara's particular sense. The three computer poems, meanwhile, exhibit other traits that Fried sees as enabling art to function "like
a surrogate person." Though none boasts the imposing physical scale of a
minimalist sculpture, they display words quickly enough to overwhelm
the senses in a different way. The visual boldness and simplicity of the
YHCHI and Lin poems, their flatness, lends them a counterintuitive force
of verbal expression, an almost kinetic energy the reader must confront.
Fried also notes the "apparent hollowness" of minimalist sculpture, the
impression that it has an inaccessible interior into which one might nonetheless tumble (156). With its story of a bottomless pit and its mesmer-
izing visual effects, Poundstone's work produces an equivalent sense of
hollowness, of approaching an abyss. Pressman argues that *Project* thereby
"becomes a bottomless pit for a reading experience that performs its message," literalizing the call for "deep" readerly attention in the figure of a deep
hole.[10] But this performative effect in *Project* also invokes its own lyric
objecthood, its power to figure both a physical topology (the pit) and a

subjective interiority of the sort Fried describes (the hollowness of the other). All three pieces thus use the computer as a platform to renegotiate the poetic expression of personhood, of subjective presence and interiority.

Finally, these computer poems convey a sense of lyric objecthood most powerfully by addressing poetry's relation to the human body, a relation most other computer poems handle poorly. The objecthood of the body provides a foundation for the primordial gestures of poetic expression: the breath and voice, the movements of a writing hand. The technologies of lyricism have developed in direct response to our ways of imagining these basic poetic gestures. It might seem counterintuitive to read the work of YHCHI, Poundstone, and Lin as meaningfully related to the human body and its gestures. After all, many other computer poems enable us to interact gesturally with the text using the keyboard or mouse, for instance by navigating a virtual space or clicking on certain words. But these poems by YHCHI, Poundstone, and Lin involve no gestural or haptic interaction whatsoever. Once they have begun, the reader can only sit back and observe. Everyday videos on the web allow the viewer to pause or skip forward with a mouse click, thus providing a richer gestural interface than these poems do. Yet these poems more powerfully invoke the objecthood of the human body precisely because they foreclose gestural interaction: through this exclusion, they underscore the fact that computers are not merely audio-visual but also gesturo-haptic media. In this way, they reflect a gestural aesthetic that Fried associates with his favorite sculptures. Even without moving, sculptures achieve gestural effects through what Fried calls their "syntax," the relations of their parts to one another (162). He views this syntactic dimension of sculpture as "imitating, not gestures exactly, but the *efficacy* of gesture" as a vehicle of aesthetic statement. By the same token, the computer poems I have been reading do not literally activate the movements of the viewer's body, but they develop their own gestural sensibilities. In *Dakota,* for instance, the syncopated interplay between the drumbeat and the flashing words has a discernibly gestural effect. Other computer poems enable the reader to interact through mouse clicks and keystrokes, but by reducing the complexity of embodied movement to these relatively simple interfaces, they lack the gestural syntax Fried describes.

Even if artworks do not move or cause us to move, they can present gesture as the human body's way to make meaning, the basis of our ability to signify. As Fried puts it, "like certain music and poetry," gestural sculptures "are possessed by the knowledge of the human body and how, in innumerable ways and moods, it makes meaning" (162). Instead of fetishizing our embodied interface with computers, the best computer poems develop their own gestural syntax, their own ways of figuring the body's objecthood. They move beyond any single technological interface in order to address the gestural basis of "meaningfulness *as such*," the boundary where the physical becomes verbal, where a mere object becomes an artwork. These poems thus reflect the shifting figural apparatus that structures the gestures by which poetry gets made and by which it gets read, the gestures by which it confronts the human body. This technological and conceptual equipment through which we think about the moving bodies that make poetry has changed a great deal since computers and electronics became common, and it will continue to change as poets respond to technological development. Further study of the lyric's afterlives, therefore, might address not only the emerging forms of computer-based poetry but also the shifting theories of poetic gesture these works convey.

Acknowledgments

I am no good at asking for advice. What I intend as a question too often comes out sounding like a confident assertion. I have talked myself right past a lot of good counsel in this way. I am therefore all the more grateful to the many friends, colleagues, and mentors who helped this book come to be, since they found ways of giving support to someone so bad at asking for it.

This project's early development benefited from a lively and encouraging community at Cornell University. I am especially indebted to my doctoral committee: Jonathan Culler, Roger Gilbert, Christopher Nealon, and Amy Villarejo. With their wisdom and generosity, they not only made this book possible but also modeled a deeply affirmative sense of scholarly vocation that guides me to this day. Other faculty at Cornell provided valuable mentorship at important junctures: thanks to Kevin Attell, Anne-Lise François, Ellis Hanson, Jenny Mann, Masha Raskolnikov, and Nick Salvato. My time in Ithaca would have been far less enjoyable and productive without the social and intellectual camaraderie I found in groups such as the Society for the Humanities, the School of Criticism and Theory, and the Theory Reading Group. I also want to thank my fellow graduate students at Cornell, who had as much a hand in the formation of these ideas as anyone, especially Chad Bennett, Owen Boynton, Jacob Brogan, Lily Cui, Stephanie DeGooyer, Bradley Depew, Ryan Dirks, Sarah Ensor, Ben Glaser, Diana Hamilton, Aaron Hodges, Jess Keiser, Rob Lehman, Anthony Reed, Cecily Swanson, and Audrey Wasser. Thank you for your patient ears and your sharp minds.

My transition from graduate school in Ithaca to a faculty position in Oklahoma accelerated my research thanks to conversations with colleagues at Oklahoma State, especially Toby Beauchamp, Jonathan Gaboury, Kate Hallemeier, Jeff Menne, Tim Murphy, and Graig Uhlin. Among my colleagues at the University of Oklahoma, I am most grateful to Daniela Garofalo, Ron Schleifer, Jonathan Stalling, and Jim Zeigler for generative conversations about my research and excellent advice about publishing it. A grant from the Oklahoma Humanities Council supported archival research for this project.

I was honored to receive the NEH Postdoctoral Fellowship in Poetics at Emory University's Fox Center for Humanistic Inquiry while I worked on this book. I am grateful to the fellows and hosts at the Fox Center for providing such a rigorous and convivial community; I could not have asked for better research conditions. I thank the National Endowment for the Humanities for helping to fund the position, one of too few designated for research in poetics.

During the later stages of writing and revision, my colleagues at Georgetown offered insightful and productive feedback. I am especially grateful to Brian Hochman, Ricardo Ortiz, Nicole Rizzuto, and Dan Shore, who read and helped me to improve various parts of the manuscript. I am grateful to Melissa Jones of the Georgetown Library for her excellent sleuthing. I thank Doug Armato and Gabe Levin at the University of Minnesota Press, as well as the anonymous reviewers who provided such thorough and astute reports. I would be remiss not to thank the member of the press's faculty advisory board who offered very constructive last-minute ideas for revision.

I owe a great deal to the broader network of poets and scholars who have given various kinds of feedback and support as this book developed, sometimes in one important conversation and sometimes through years of exchange. I received excellent advice on the third chapter during my visit to the Poetry and Poetics Workshop at the University of Chicago; thanks especially to Rachel Galvin, Eric Powell, Chicu Reddy, Lauren Schachter, and John Wilkinson. Part of the first chapter was published in *Criticism*, part of the third in *Paideuma*; thanks to both journals for their support. I am grateful to Jennifer Ashton, Steve Benson, Bill Berkson, Charles Bernstein, Mutlu Blasing, Bob Cataliotti, Michael Clune, Johanna Drucker, Jonathan Eburne, Christopher Funkhouser, Maureen O'Hara Granville-Smith, Billy Joe Harris, Charles O. Hartman, Katherine Hayles,

Oren Izenberg, Lauren Klein, David Nowell Smith, Yasmine Shamma, Lytle Shaw, Martha Nell Smith, Michael Snediker, Alexandra Socarides, Anne Tardos, Johanna Winant, Aaron Winslow, Kevin Young, and Steven Zultanski. My greatest debts in this project, as in all things, are to my family. I owe them immeasurable gratitude for their encouragement, warmth, and support through the years. Thank you to my parents, David and Joan Perlow, to whom this book is dedicated; to Sam and Dara Matthew; and to Marjorie Stevens. You make everything feel possible. Finally, thank you to the four animals who populate my every day. The vocal cat Baker and the two Shih Tzu, Frisco and Jimmy Carter, have remained wordlessly oblivious to this entire effort, keeping it in proper perspective. As to human companionship, Caetlin Benson-Allott, you are the most fantastically wonderful partner I can imagine. If you do not know the value of all you do to inspire and sustain me, it is written across every page of this book. To share work and life with you is my greatest privilege and pleasure.

Notes

INTRODUCTION

1. William Carlos Williams, *The Wedge* (Cummington, Mass.: Cummington Press, 1944), 8.

2. William Carlos Williams, *Spring and All* (Paris: Contact, 1923), 92.

3. Plato's *Ion* deploys an electrical metaphor to distinguish poetry from knowledge when Socrates compares the poet to an iron ring dangling from a magnet, the source of his inspiration (533c–e and 535e–36e, trans. Paul Woodruff in *Plato: Complete Works*, ed. John M. Cooper [Indianapolis, Ind.: Hacketty, 1997], 941 and 943–44, respectively).

4. Walt Whitman, *Leaves of Grass* (New York: W. E. Chapin, 1867), 98, 21.

5. The term "information technology" emerged contemporaneously with Williams's maxims about poetry and machines. According to the Oxford English Dictionary, it first appeared in Merrill M. Flood's report on David Rosenblatt's "Aspects of Technology and Organization," a paper the latter gave at the 1951 meeting of the Econometric Society. Here the term refers to mathematical techniques, not electronics. Flood already confirms the importance of information's others when he emphasizes the "nonmathematical" factors econometrics must take into account. He was then an employee of the RAND Corporation, discussed at length in chapter 2 (see "Report of the Boston Meeting," *Econometrica: Journal of the Econometric Society* 20, no. 3 [July 1952]: 499–500).

6. Craig Dworkin, *Reading the Illegible* (Evanston, Ill.: Northwestern University Press, 2003); Virginia Jackson, *Dickinson's Misery: A Theory of Lyric Reading* (Princeton, N.J.: Princeton University Press, 2005); Steve McCaffery, *The Darkness of the Present: Poetics, Anachronism, and the Anomaly* (Tuscaloosa: University of Alabama Press, 2012); Daniel Tiffany, *Infidel Poetics: Riddles, Nightlife, Substance* (Chicago: University of Chicago Press, 2009).

7. In *Republic* 10.601.a, Plato writes: "A poetic imitator uses words . . . to paint colored pictures of each of the crafts. He himself knows nothing about them, but . . .

others, as ignorant as he, who judge by words, will think he speaks extremely well" (trans. G. M. A. Grube and rev. C. D. C. Reeve in *Complete Works*, 1205). Plato comments further on imitative poetry in *Ion* (535a–36e; *Complete Works*, 942–44) and on writing and memory in *Phaedrus* (274e–75b; *Complete Works*, 551–52). Both treatments express doubt about writing's truthfulness, and both offer technological figures for writing, one of which involves magnetism, an electrical phenomenon.

8. Thomas Gray, "Ode on a Distant Prospect of Eton College," *The Thomas Gray Archive: A Collaborative Digital Collection*, accessed July 14, 2017, http://www.thomas gray.org/cgi-bin/display.cgi?text=odec. Many have noted Gray's fondness for Plato, but few expound upon it. Perhaps the most useful account of his classical studies is LaRue van Hook, "New Light on the Classical Scholarship of Thomas Gray," *The American Journal of Philology* 57, no. 1 (1936): 1–9.

9. William Wordsworth, "Ode [Intimations of Immortality]," ll.58–76, in *William Wordsworth*, ed. Stephen Gill, 21st-Century Oxford Authors (New York: Oxford University Press, 2010), 282. Gill cites a note from Wordsworth linking the first of these lines with "Platonic philosophy" (764).

10. Whitman, *Leaves of Grass*, 28.

11. E. E. Cummings, "53," in *100 Selected Poems* (New York: Grove Press, 1954), 65.

12. Wallace Stevens, "Notes toward a Supreme Fiction," *The Collected Poems* (New York: Vintage, 1990), 380.

13. This perspective on Stevens contradicts scholars' preference for the long "philosophical" poems written later in his career, but as I have argued elsewhere, anthology editors and casual readers get something right in their preference for his shorter, earlier, more elliptical poems; see Seth Perlow, "The Other *Harmonium*: Toward a Minor Stevens," *The Wallace Stevens Journal* 33, no. 2 (Fall 2009): 191–210.

14. Allen Grossman, "Summa Lyrica: A Primer of the Commonplaces in Speculative Poetics," in *The Sighted Singer: Two Works on Poetry for Readers and Writers* (Baltimore, Md.: Johns Hopkins University Press, 1992), 209.

15. Grossman, "Summa Lyrica," 247.

16. Gillian White tracks these anxieties about the embrace and rejection of lyric norms in recent American poetry, arguing that lyricism now arouses shame (*Lyric Shame: The "Lyric" Subject of Contemporary American Poetry* [Cambridge, Mass.: Harvard University Press, 2014]).

17. Dworkin and Goldsmith make this shared opposition explicit in their introductions to their coedited *Against Expression: An Anthology of Conceptual Writing* (Evanston, Ill.: Northwestern University Press, 2011).

18. Mutlu Konuk Blasing, *Lyric Poetry: The Pain and the Pleasure of Words* (Princeton, N.J.: Princeton University Press, 2007), 10.

19. Virginia Jackson develops this critique most decisively in *Dickinson's Misery*. See also Jackson et al., "The New Lyric Studies," *PMLA* 123, no. 1 (January 2008): 181–234, and *The Lyric Theory Reader*, ed. Virginia Jackson and Yopie Prins (Baltimore, Md.: Johns Hopkins University Press, 2013).

20. Fredric Jameson, *The Political Unconscious: Narrative as a Socially Symbolic Act* (Ithaca, N.Y.: Cornell University Press, 1982), 9.

21. These terms of division circulate in common, with the exception of "cooked/raw," which Robert Lowell proposed in a speech after receiving the National Book Award for *Life Studies* in 1960 (Ian Hamilton, *Robert Lowell: A Biography* [New York: Random House, 1982], 277). Many scholars trace this division to the competition between two anthologies, the relatively traditional *New Poets of England and America*, ed. Donald Hall, Robert Pack, and Louis Simpson (New York: Meridian, 1957), and the more experimental *The New American Poetry 1945–1960*, ed. Donald Allen (New York: Grove Press, 1960), which published Allen Ginsberg, Frank O'Hara, Jack Spicer, Charles Olson, and other experimentalists mentioned in this study.

22. See, for instance, the conciliatory but tepid forewords to *American Hybrid: A Norton Anthology of New Poetry*, ed. Cole Swensen and David St. John (New York: W. W. Norton, 2009).

23. See: Jackson, *Dickinson's Misery*; Helen Vendler, *Our Secret Discipline: Yeats and Lyric Form* (New York: Oxford University Press, 2007); Robert von Hallberg, *Lyric Powers* (Chicago: University of Chicago Press, 2008); Michael Snediker, *Queer Optimism: Lyric Personhood and Other Felicitous Persuasions* (Minneapolis: University of Minnesota Press, 2008); Jackson and Prins, *The Lyric Theory Reader*; White, *Lyric Shame*; and Hannah Vandegrift Eldridge, *Lyric Orientations: Hölderlin, Rilke, and the Poetics of Community* (Ithaca, N.Y.: Cornell University Press, 2016).

24. In *The Art of Shakespeare's Sonnets* (Cambridge, Mass.: Harvard University Press, 1997), Helen Vendler offers an alternative to the opposition between expressive and mimetic poetics when she reads expressive poetry as mimetic of a subjective interior. Among others, Jonathan Culler claims this move reduces lyric poems to the condition of dramatic monologues, theatrically staging selfhood but ignoring the lyric's ritualistic dimensions (*Theory of the Lyric* [Cambridge, Mass.: Harvard University Press, 2015], 244–74).

25. Marjorie Perloff, *21st-Century Modernism: The "New" Poetics* (New York: Blackwell, 2002), 158.

26. Virginia Jackson, "Lyric," *The Princeton Encyclopedia of Poetry and Poetics*, ed. Roland Greene et al., 4th ed. (Princeton, N.J.: Princeton University Press, 2012), 826.

27. Nikki Skillman, *The Lyric in the Age of the Brain* (Cambridge, Mass.: Harvard University Press, 2016), 4.

28. Skillman also persuasively questions whether expression really operates as the key term of division between avant-garde and conventional American poets today (*The Lyric in the Age of the Brain*, 17–18).

29. Von Hallberg, *Lyric Powers*, 11, 1.

30. Blasing, *Lyric Poetry*, 4, 1–2.

31. Jackson, "Lyric," 826.

32. C. P. Snow, *The Two Cultures and the Scientific Revolution: The Rede Lecture* (New York: Cambridge University Press, 1959). Twenty-five years later, Thomas Pynchon

confirms the technocratic victory Snow foretells: "Writers of all descriptions are stampeding to buy word processors. Machines have already become so user-friendly that even the most unreconstructed Luddites can be charmed into laying down the old sledgehammer and stroking a few keys instead" ("Is It O.K. To Be A Luddite?" *New York Times,* Books, October 28, 1984, 41).

33. Northrop Frye is not alone in considering lyric "the genre which most clearly shows the hypothetical core of literature" (*Anatomy of Criticism: Four Essays* [Princeton, N.J.: Princeton University Press, 1957], 271). See my discussion of the lyric's exemplarity below.

34. See Peter Middleton, *Physics Envy: American Poetry and Science in the Cold War and After* (Chicago: University of Chicago Press, 2015). Reading postwar and contemporary American poets, Middleton diagnoses "physics envy," an impulse to claim for poetry some of the cultural prestige and epistemic authority of the physical sciences. This account sees experimental poets as "reaching toward potentially new positive knowledge," an impulse precisely opposite the rhetoric of lyric exemption, but Middleton also addresses "poets for whom physics envy . . . encourages them to think harder about epistemological questions," a more distanced response to technoscientific models for poetry (2, 11).

35. Some scholarship that avoids equating electronics with rationalism nonetheless suggests that a poet's experiments with electronics will disrupt rather than affirm lyric norms; see Loss Pequeño Glazier, *Digital Poetics: The Making of E-Poetries* (Tuscaloosa: University of Alabama Press, 2002). Other studies of poetry that attend to the less rational functions of electronics include Lori Emerson, *Reading Writing Interfaces: From the Digital to the Bookbound* (Minneapolis: University of Minnesota Press, 2014), and R. John Williams, *The Buddha in the Machine: Art, Technology, and the Meeting of East and West* (New Haven, Conn.: Yale University Press, 2014). A few studies link electronic media with lyricism in particular, such as: Carrie Noland, *Poetry at Stake: Lyric Aesthetics and the Challenge of Technology* (Princeton, N.J.: Princeton University Press, 1999); Brian Kim Stefans, *Word Toys: Poetry and Technics* (Tuscaloosa: University of Alabama Press, 2017); and Daniel Tiffany, *Toy Medium: Materialism and Modern Lyric* (Berkeley: University of California Press, 2000).

36. Influential studies emphasizing the informational functions of literary technologies include Marjorie Perloff's *Radical Artifice: Writing Poetry in the Age of Media* (Chicago: University of Chicago Press, 1991), a foundational work in media-oriented poetics, and N. Katherine Hayles's *Electronic Literature: New Horizons for the Literary* (Notre Dame, Ind.: University of Notre Dame Press, 2008) and *How We Think: Digital Media and Contemporary Technogenesis* (Chicago: University of Chicago Press, 2012), which argue that literature has become computational. Some who recognize that poetry can disrupt technorationalist frameworks continue to associate electronics with information and scientific knowledge; see Paul Stephens's excellent *The Poetics of Information Overload: From Gertrude Stein to Conceptual Writing* (Minneapolis: University of Minnesota Press, 2015), which affirms that electronics produce information

and tracks poetic responses to overwhelming amounts of it, and Christian Bök's *Pataphysics: The Poetics of an Imaginary Science* (Evanston, Ill.: Northwestern University Press, 2002), which develops a poetic history of a pararational science that defines itself in relation to dominant technoscientific modes of knowledge.

37. Recent scholarship on poetry of this kind includes Brian M. Reed, *Nobody's Business: Twenty-First Century Avant-Garde Poetics* (Ithaca, N.Y.: Cornell University Press, 2013), and Daniel Morris, *Not Born Digital: Poetics, Print Literacy, New Media* (New York: Bloomsbury, 2016).

38. See David L. Hoover, "Argument, Evidence, and the Limits of Digital Literary Studies," and Joanna Swafford, "Messy Data and Faulty Tools," both in *Debates in the Digital Humanities 2016*, ed. Matthew K. Gold and Lauren F. Klein (Minneapolis: University of Minnesota Press, 2016), 230–50 and 556–58, respectively. The seminal examples of quantitative digital humanities scholarship are Franco Moretti's *Distant Reading* (New York: Verso, 2013) and *Graphs, Maps, Trees: Abstract Models for Literary History* (New York: Verso, 2007). As Moretti uses "big data" to track the rise and fall of fiction genres, for instance, he makes little effort to theorize the parameters by which his data differentiate one genre from another, and even less to consider the more fundamental and far more challenging question of what constitutes a genre in the first place.

39. Media archaeologists have addressed these more complex functions of reading and writing equipment, though generally without addressing how such matters inform literary criticism. See: Johanna Drucker, *Graphesis: Visual Forms of Knowledge Production* (Cambridge, Mass.: Harvard University Press, 2014); Juliet Fleming, *Cultural Graphology: Writing after Derrida* (Chicago: University of Chicago Press, 2016); Lisa Gitelman, *Paper Knowledge: Toward a Media History of Documents* (Durham, N.C.: Duke University Press, 2014); and Matthew G. Kirschenbaum, *Mechanisms: New Media and the Forensic Imagination* (Cambridge, Mass.: MIT Press, 2012).

40. See: *Emily Dickinson Archive,* edickinson.org; *The Dickinson Electronic Archives,* emilydickinson.org; *Thomas Gray Archive,* thomasgray.org; *Rossetti Archive,* rossetti archive.org; and *The Walt Whitman Archive,* whitmanarchive.org.

41. Exceptions to this norm include Natalie M. Houston, "Toward a Computational Analysis of Victorian Poetics," *Victorian Studies* 56, no. 3 (Spring 2014): 498–510, and the Trans-Historical Poetry Project at the Stanford Literary Lab, litlab.stanford.edu.

42. Wendy Hui Kyong Chun, Richard Grusin, Patrick Jagoda, and Rita Raley, "The Dark Side of the Digital Humanities," in Gold and Klein, *Debates in the Digital Humanities 2016*, 493. For a more scathing assessment, see Daniel Allington, Sarah Brouillette, and David Golumbia, "Neoliberal Tools (and Archives): A Political History of Digital Humanities," *Los Angeles Review of Books*, May 1, 2016, lareviewofbooks.org/article/neoliberal-tools-archives-political-history-digital-humanities.

43. See, for instance, Paul Virilio's account of "disappearance" (*The Aesthetics of Disappearance,* trans. Phil Beitchman [New York: Semiotext(e), 2009; originally 1980]) and Bernard Stiegler's theory of "disorientation" (*Technics and Time, 2: Disorientation,*

trans. Stephen Barker [Stanford, Calif.: Stanford University Press, 2009; originally 1996]), terms for technology's power to disrupt our perceptive and rational capacities. Though Virilio and Stiegler differ in significant ways, they both develop useful vocabularies for the less rational functions of electronics, and both offer negative judgments about these effects.

44. The first use of "bit" is contested. Shannon, who describes it as a portmanteau of "binary" and "digit," attributes the term to J. W. Tukey (Shannon, "A Mathematical Theory of Communication," *Bell System Technical Journal* 27, no. 3 [July 1948]: 379–423), but as early as 1936, Vannevar Bush referred to "bits of information" on a punch card ("Instrumental Analysis," *Bulletin of the American Mathematical Society* 42, no. 10 [1936]: 649–69).

45. A. M. Turing, "Computing Machinery and Intelligence," *Mind: A Quarterly Review of Psychology and Philosophy* 59, no. 236 (October 1950): 433–60.

46. Turing writes that "we also wish to allow the possibility that an engineer or team of engineers may construct a machine which works, but whose manner of operation cannot be satisfactorily described by its constructors because they have applied a method which is largely experimental," but he does "exclude ... men born in the usual manner," a potential subset of machines whose creators do not understand their inner workings ("Computing Machinery and Intelligence," 435).

47. Gilbert Simondon views technology's reliance upon humans for maintenance and intermediation as a sign of human freedom, a limit on mechanical determination that also grants us a critical perspective on the influences of technology: "The action of having to intervene as a mediator in this relation between machines grants [man] a situation of independence in which he can acquire a cultural vision of technical realities" (*On the Mode of Existence of Technical Objects,* trans. Cecile Malaspina and John Rogove [Minneapolis: Univocal, 2017; originally 1958], 159).

48. In addition to above mentioned studies by Emerson, Noland, Stefans, and Tiffany, see: Susan McCabe, *Cinematic Modernism: Modernist Poetry and Film* (New York: Cambridge University Press, 2005); David Trotter, *Literature in the First Media Age: Britain between the Wars* (Cambridge, Mass.: Harvard University Press, 2013); and Christophe Wall-Romana, *Cinepoetry: Imaginary Cinemas in French Poetry* (New York: Fordham University Press, 2013).

49. The best introductions to such work are C. T. Funkhouser's *Prehistoric Digital Poetry: An Archaeology of Forms, 1959–1995* (Tuscaloosa: University of Alabama Press, 2007) and *New Directions in Digital Poetry* (New York: Continuum, 2012) and Glazier's *Digital Poetics.*

50. On the word processor, see Matthew G. Kirschenbaum, *Track Changes: A Literary History of Word Processing* (Cambridge, Mass.: Harvard University Press, 2016).

51. For instance, see Michel de Certeau, *The Practice of Everyday Life,* trans. Steven Rendall (Berkeley: University of California Press, 1984; originally 1980), and Henri Lefebvre, *Everyday Life in the Modern World,* trans. Sacha Rabinovitch (New York: Harper & Row, 1971; originally 1968).

52. Simondon, *On the Mode of Existence of Technical Objects*, 18.

53. Cf. Martin Heidegger, *Being and Time*, trans. John Macquarrie and Edward Robinson (New York: Blackwell, 1962), 95–101.

54. Heidegger, *Being and Time*, 96.

55. Heidegger, *Being and Time*, 97. Heidegger affiliates "equipment" with the Greek *pragmata*. The more familiar term for this sense of usefulness is "ready-to-hand," but *Zeug* can also mean "stuff" in the collective, and sometimes pejorative, sense.

56. Heidegger, *Being and Time*, 97.

57. On Heidegger's theory of technology, see Don Ihde, *Heidegger's Technologies: Postphenomenological Perspectives* (New York: Fordham University Press, 2010).

58. Martin Heidegger, "The Question Concerning Technology," trans. William Lovitt, in *Basic Writings*, ed. David Farrell Krell (New York: Harper Collins, 1993), 318.

59. On this ambivalence in Heidegger's theory of technology, see Ihde, *Heidegger's Technologies*, 28–35.

60. On Heidegger's theory of poetry, see David Nowell Smith, *Sounding/Silence: Martin Heidegger at the Limits of Poetics* (New York: Fordham University Press, 2013), and Krzysztof Ziarek, *Language after Heidegger* (Bloomington: Indiana University Press, 2013), 130–74.

61. Highly quantitative, informational methods not only shape foundational DH studies, such as Moretti's *Graphs, Maps, Trees*, but also remain the predominant methodological framework for more recent work in the field, including several of the contributions to Gold and Klein, *Debates in the Digital Humanities 2016*.

62. See Emerson, *Reading Writing Interfaces*; Noland, *Poetry at Stake*; Jessica Pressman, *Digital Modernism: Making It New in New Media* (New York: Oxford University Press, 2014); Stefans, *Word Toys*; and Stephens, *The Poetics of Information Overload*.

63. René Descartes, *Meditations on First Philosophy*, ed. Stanley Tweyman, trans. Elizabeth S. Haldane and G. R. T. Ross (New York: Routledge, 1993), 45.

64. For this reason, Jürgen Habermas calls Cartesian doubt "a performative contradiction" (Jürgen Habermas, *On the Pragmatics of Communication*, ed. Maeve Cooke [Cambridge, Mass.: MIT Press, 1998], 356), an insight he credits to Charles Peirce and Ludwig Wittgenstein. The latter remarks: "If you are not certain of any fact, you cannot be certain of the meaning of your words either. . . . If you tried to doubt everything you would not get as far as doubting anything" (*On Certainty*, trans. Denis Paul and G. E. M. Anscombe, ed. G. E. M. Anscombe and G. H. von Wright [Oxford: Basil Blackwell, 1969], §§ 114–15).

65. Descartes, *Meditations on First Philosophy*, 41.

66. Edmund Husserl, *Cartesian Meditations: An Introduction to Phenomenology*, trans. Dorion Cairns (Dordrecht, The Netherlands: Kluwer Academic, 1999), 2.

67. A different, absolutist rhetoric against knowledge, equally important for philosophers if not for poets, imagines everything beyond the closure of rationalism as rigorously unavailable to the understanding. This line goes back to the discourse of

Platonic forms, but its most influential recent exemplars are the negative theologies of Emmanuel Levinas and his followers, notably Jean-Luc Marion and Michel Henry. These exponents of an "ethical turn" in phenomenology imagine the ethical relation as an encounter with a radically unknowable other; their more or less explicitly theological appeals to infinitude entail a rightist exclusion of plurality and pragmatic politics; see Dominique Janicaud et al., *Phenomenology and the "Theological Turn": The French Debate* (New York: Fordham University Press, 2000), and J. Aaron Simmons and Bruce Ellis Benson, *The New Phenomenology: A Philosophical Introduction* (New York: Bloomsbury Academic, 2013).

68. In *Lyric Poetry,* Blasing deploys a similar figure-ground trope to cast poetic forms as prior and exterior to rational discourse but also able to host the latter: "Poetic forms clearly accommodate referential use of language and rational discourse. But they position most complex thought processes and rigorous figurative logic as figures on the ground of processes that are in no way rational" (3). By these lights, the poetic moment consists not in extrarational discourse per se, but in the rhetoric that produces a sense of poetic form as prior to the rational figures it might host.

69. *Oxford English Dictionary Online,* "exempt."

70. *Oxford English Dictionary Online,* "example."

71. On the rhetoric and logic of exemplarity, see *Exemplarity and Singularity: Thinking through Particularity in Philosophy, Literature, and Law,* ed. Michèle Lowrie and Susanne Lüdemann (New York: Routledge, 2015).

72. I borrow "queer epistemology" from David L. Eng, Judith Halberstam, and José Esteban Muñoz, "Introduction: What's Queer about Queer Studies Now?" *Social Text* 23, nos. 3–4 (84–85) (Fall–Winter 2005): 1–17.

73. Eve Kosofsky Sedgwick, "Paranoid Reading and Reparative Reading, or, You're So Paranoid, You Probably Think This Essay Is About You," in *Touching Feeling: Affect, Pedagogy, Performativity* (Durham, N.C.: Duke University Press, 2003), 125, 143. In her earlier work, Sedgwick argues that different kinds of ignorance "are produced by and correspond to particular knowledges and circulate as part of particular regimes of truth," another structure of mutual dependence between knowledge and its others ("Privilege of Unknowing: Diderot's *The Nun,*" in *Tendencies* [Durham, N.C.: Duke University Press, 1993], 25).

74. A wide variety of recent humanities scholarship proposes alternatives to rationalist norms. For two works that directly and influentially address the question of rational criticism's limits, see Rita Felski, *The Limits of Critique* (Chicago: University of Chicago Press, 2015), and Bruno Latour, "Why Has Critique Run Out of Steam? From Matters of Fact to Matters of Concern," *Critical Inquiry* 30 (Winter 2004): 225–48.

75. *Oxford English Dictionary Online,* "mode."

76. Susan Howe, "These Flames and Generosities of the Heart: Emily Dickinson and the Illogic of Sumptuary Values," in *The Birth-mark: Unsettling the Wilderness in American Literary History* (Middletown, Conn.: Wesleyan University Press, 1993), 136.

77. Marta L. Werner, "The Flights of A 821: Dearchivizing the Proceedings of a Birdsong," in *Voice, Text, Hypertext: Emerging Practices in Textual Studies,* ed. Raimonda Modiano, Leroy F. Searle, and Peter Shillingsburg (Seattle: University of Washington Press, 2004), 305.

78. Jackson Mac Low, "The Poetics of Chance and the Politics of Simultaneous Spontaneity, or the Sacred Heart of Jesus (Revised & Abridged) July 12, 1975," in *Talking Poetics from the Naropa Institute,* ed. Anne Waldman and Marilyn Webb (Boulder, Colo.: Shambhala, 1978), 176; Mac Low, "Museletter," in *The L=A=N=G=U= A=G=E Book,* ed. Bruce Andrews and Charles Bernstein (Carbondale: Southern Illinois University Press, 1984), 26.

79. Larry Rivers, "Speech Read at Frank O'Hara's Funeral," in *Homage to Frank O'Hara,* ed. Bill Berkson and Joe LeSeur (Bolinas, Calif.: Big Sky, 1988), 138.

80. Frank O'Hara, "Morning," in *The Collected Poems of Frank O'Hara,* ed. Donald Allen (Berkeley: University of California Press, 1995), 32.

81. Frank O'Hara, "Personism: A Manifesto," in *Collected Poems,* 499.

82. See Daniel Belgrad, *The Culture of Spontaneity: Improvisation and the Arts in Postwar America* (Chicago: University of Chicago Press, 1998).

1. AFFECT

1. An earlier possession of Dickinson appeared in the lyrics of a Simon & Garfunkel tune: "And you read your Emily Dickinson / And I my Robert Frost" ("The Dangling Conversation," from *Parsley, Sage, Rosemary and Thyme* [Columbia, 1966]). Susan Howe makes clear that her possessive formula comes directly from Dickinson, but Simon & Garfunkel's gendered organization of possession proves resonant.

2. Jack Spicer, "The Poems of Emily Dickinson," in *The House That Jack Built: The Collected Lectures of Jack Spicer,* ed. Peter Gizzi (Middletown, Conn.: Wesleyan University Press, 1998), 232.

3. Susan Howe, "These Flames and Generosities of the Heart: Emily Dickinson and the Illogic of Sumptuary Values," in *The Birth-mark: Unsettling the Wilderness in American Literary History* (Middletown, Conn.: Wesleyan University Press, 1993), 136.

4. On the most influential theories of affect and their increasing significance for humanities scholars during the past half century, see Ruth Leys, *The Ascent of Affect: Genealogy and Critique* (Chicago: University of Chicago Press, 2017).

5. In *Our Emily Dickinson: American Women Poets and the Intimacies of Difference* (Philadelphia: University of Pennsylvania Press, 2017), Vivian R. Pollak explores the affective and gender dynamics of Dickinson's reception among an earlier generation of poets, including Bishop, Moore, Plath, Rich, and Rukeyser.

6. Philip F. Gura, "How I Met and Dated Miss Emily Dickinson: An Adventure on eBay," *Common-place* 4, no. 2 (January 2004), common-place.org/book/how-i-met -and-dated-miss-emily-dickinson-an-adventure-on-ebay/.

7. George Gleason, "Is It Really Emily Dickinson?," *The Emily Dickinson Journal* 18, no. 2 (Fall 2009): 3.

8. Susan M. Pepin, "Comparison of Emily Dickinson's orbit/eyelid anatomy from the daguerreotype of 1847 and the discovered daguerreotype of two women of 1859," *Dickinson Electronic Archives,* October 7, 2012, emilydickinson.org/node/12.

9. Walter Benjamin, "The Work of Art in the Age of Mechanical Reproduction," in *Illuminations,* ed. Hannah Arendt, trans. Harry Zohn (New York: Schocken Books, 1968), 226.

10. Martha Nell Smith, "Iconic Power and the New Daguerreotype of Emily Dickinson," *1859 Daguerreotype: Is This Emily Dickinson,* Emily Dickinson Archive, emily dickinson.org/node/20.

11. Smith, "Iconic Power."

12. Millicent Todd Bingham, *Ancestors' Brocades: The Literary Debut of Emily Dickinson* (New York: Harper, 1945), 224; Smith, "Iconic Power."

13. Ruth Leys, "The Turn to Affect: A Critique," *Critical Inquiry* 37 (Spring 2011): 438. Leys paraphrases this belief in order to critique it.

14. Michael D. Snediker, *Queer Optimism: Lyric Personhood and Other Felicitous Persuasions* (Minneapolis: University of Minnesota Press, 2009), 99, 109.

15. Emily Dickinson, *The Poems of Emily Dickinson: Variorum Edition,* ed. R. W. Franklin (Cambridge, Mass.: Harvard University Press, 1998), 339. Poems from this edition are hereafter cited in the text as F followed by poem number. Except where otherwise noted, the first variant is cited.

16. See Silvan Tomkins, *Affect Imagery Consciousness,* 2 vols. (New York: Springer, 1962–1963); *Emotion in the Human Face,* ed. Paul Eckman (New York: Pergamon, 1972).

17. Joan Kirby, "'[W]e thought Darwin had thrown "the redeemer" away': Darwinizing with Emily Dickinson," *The Emily Dickinson Journal* 19, no. 1 (2010): 11–12.

18. See Virginia Jackson, *Dickinson's Misery: A Theory of Lyric Reading* (Princeton, N.J.: Princeton University Press, 2005).

19. Emily Dickinson, *The Letters of Emily Dickinson,* ed. Thomas H. Johnson (Cambridge, Mass.: Harvard University Press, 1965), 710. Hereafter, this work will be cited in the text as L.

20. Don Gilliland analyzes Dickinson's religious faith in relation to her poetic production, focusing upon the idea of "uncertain certainty." He underscores the link between faith and nonknowledge in her thought, citing her avowal, in a letter to Susan Dickinson, that "Faith is *Doubt*" as evidence of her theological sophistication (Don Gilliland, "Textual Scruples and Dickinson's 'Uncertain Certainty,'" *The Emily Dickinson Journal* 18, no. 2 [Fall 2009]: 59n27).

21. Jackson, *Dickinson's Misery,* 13.

22. Shawn Alfrey, "The Function of Dickinson at the Present Time," *The Emily Dickinson Journal* 11, no. 1 (Spring 2002): 11.

23. Robert McClure Smith, "Dickinson and the Masochistic Aesthetic," *The Emily Dickinson Journal* 7, no. 2 (Fall 1998): 16.

24. W. H. Auden, "In Memory of W. B. Yeats," *The New Republic* 98, no. 1266 (March 8, 1939): 123.

25. See Billy Collins, "Taking Off Emily Dickinson's Clothes," *Harper's Magazine* 296, no. 1776 (May 1998): 30–31.

26. Susan Stewart, "Lyric Possession," *Critical Inquiry* 22, no. 1 (Autumn 1995): 34.

27. Susan Stewart, *Poetry and the Fate of the Senses* (Chicago: University of Chicago Press, 2002), 143.

28. The most instructive philosophies of affect in relation to subject formation come from readers of Edmund Husserl. See Michel Henry, *Material Phenomenology,* trans. Scott Davidson (New York: Fordham University Press, 2008), who extends the line of French phenomenology, and Rei Terada, *Feeling in Theory: Emotion after the "Death of the Subject"* (Cambridge, Mass.: Harvard University Press, 2001), who addresses affect in conversation with Husserl's deconstructionist commentators.

29. Jackson, *Dickinson's Misery,* 10.

30. See R. W. Franklin, *The Editing of Emily Dickinson: A Reconsideration* (Madison: University of Wisconsin Press, 1967).

31. Harvard University Press claims to control rights to all texts written by Emily Dickinson, including those derived from manuscripts Harvard does not own. Harvard also controls permissions of images of manuscripts it owns and collects royalties from those wishing to publish them. Amherst College, the other major repository of Dickinson manuscripts, claims no rights to the texts and does not require permissions or royalties for reproduction of their manuscript images.

32. In addition to the authoritative *Emily Dickinson Archive,* notable online archives include: the *Dickinson Electronic Archives,* directed by Martha Nell Smith; *Radical Scatters,* a collection of late fragments edited by Marta Werner; *Emily Dickinson's Correspondences,* edited by Smith and Lara Vetter; and the Boston Public Library's Flickr collection of letters sent to T. W. Higginson.

33. Jerome McGann joins Cameron among notable studies of Dickinson's visible manuscripts during the early 1990s, and he offers praise for Howe's work (*Black Riders: The Visible Language of Modernism* [Princeton, N.J.: Princeton University Press, 1993]).

34. Alexandra Socarides, *Dickinson Unbound: Paper, Process, Poetics* (New York: Oxford University Press, 2012), 15, 19.

35. Marta L. Werner, "The Flights of A 821: Dearchivizing the Proceedings of a Birdsong," in *Voice, Text, Hypertext: Emerging Practices in Textual Studies,* ed. Raimonda Modiano, Leroy F. Searle, and Peter Shillingsburg (Seattle: University of Washington Press, 2004), 305.

36. Susan Howe, *My Emily Dickinson* (New York: New Directions, 2007, originally 1985), 14.

37. Howe, *My Emily Dickinson,* 51.

38. Howe, *My Emily Dickinson,* 24.

39. Spicer, "The Poems of Emily Dickinson," 231.

40. Howe, *My Emily Dickinson,* 35n1.

41. Howe, *My Emily Dickinson,* 35n1.

42. Howe, *My Emily Dickinson,* 35.

43. Howe, "These Flames," 142.

44. Quoted in Howe, "These Flames," 142.

45. Howe, "These Flames," 142.

46. Walter Benn Michaels, *The Shape of the Signifier: 1967 to the End of History* (Princeton, N.J.: Princeton University Press, 2004), 3–4.

47. Howe, "These Flames," 184.

48. Michaels, *The Shape of the Signifier,* 7.

49. Michaels, *The Shape of the Signifier,* 5.

50. Quoted in Michaels, *The Shape of the Signifier,* 9.

51. New Directions publishers bills *The Gorgeous Nothings* as "the first full-color facsimile edition of Emily Dickinson's manuscripts ever to appear," but color holograph images had appeared online over a decade earlier. Further, the *Emily Dickinson Archive,* the largest online collection of holograph images (and all in color), happens to have opened only six days before the publication of *The Gorgeous Nothings.* The latter thus elides a digital history of color facsimiles as it appropriates this strategy on behalf of an earlier medium, the printed codex. See Emily Dickinson, *The Gorgeous Nothings: Emily Dickinson's Envelope Poems,* ed. Jen Bervin and Marta Werner (New York: New Directions, 2013).

52. Howe, *My Emily Dickinson,* 130.

53. Werner later does the same in *Emily Dickinson's Open Folios* (1996), calling them "diplomatic transcriptions" (*Emily Dickinson's Open Folios: Scenes of Reading, Surfaces of Writing* [Ann Arbor: University of Michigan Press, 1996], 6). This technique might more accurately be called type facsimile, since it relies upon visual mimicry rather than editorial symbols; see Gregory A. Pass, *Descriptive Cataloging of Ancient, Medieval, Renaissance, and Early Modern Manuscripts,* Bibliographic Standards Committee, Rare Books and Manuscripts Section, Association of College and Research Libraries (Chicago: American Library Association, 2003), 144–45.

54. Howe, "These Flames," 136.

55. Howe, "These Flames," 143.

56. Howe, "These Flames," 140. In his 1955 variorum, Johnson too cites the holographs' ambiguity between verse and prose, linking it with the affective intensity of Dickinson's writing: "It is impossible in such a *jeu d'esprit* to be sure where the prose leaves off and the poetry begins." (quoted in Spicer, "The Poems of Emily Dickinson," 232n2).

57. Jerome McGann, "Emily Dickinson's Visible Language," *The Emily Dickinson Journal* 2, no. 2 (Fall 1993): 51, 42, 41.

58. Spicer, "The Poems of Emily Dickinson," 235.

59. Charles Olson, "Projective Verse," in *Collected Prose,* ed. Donald Allen and Benjamin Friedlander (Berkeley: University of California Press, 1997), 245.

60. See: Susan Howe, "An Open Field: Susan Howe in Conversation," *Academy of American Poets*, September 7, 2011, poets.org/poetsorg/text/open-field-susan-howe -conversation; and Howe, "Susan Howe, The Art of Poetry No. 97, Interviewed by Maureen N. McLane," *Paris Review* 203, Winter 2012, theparisreview.org/interviews/ 6189/susan-howe-the-art-of-poetry-no-97-susan-howe.

61. Howe, "An Open Field."

62. Karen Jackson Ford, *Gender and the Poetics of Excess: Moments of Brocade* (Jackson: University of Mississippi Press, 1997), 37.

63. In Dickinson's era, American ballet was relatively undeveloped, and the ballerina condensed anxieties about nationality as much as gender. Most skilled ballerinas were imported from Europe—from "abroad," to borrow a word from the poem— occasioning concerns not only about the gauzy "excess" of the stage props that Ford emphasizes but also about the perceived hypersexuality of their dances, the Continental influences these represented, and the young nation's inability to impart "Ballet Knowledge" to its native dancers; see Christopher Martin, "Naked Females and Splay-Footed Sprawlers: Ballerinas on the Stage in Jacksonian America," *Theatre Survey* 51, no. 1 (May 2010): 95–114.

64. Simon Jarvis, "Musical Thinking: Hegel and the Phenomenology of Prosody," *Paragraph* 28, no. 2 (July 2005): 69.

65. Such a prospect may point to a specifically national community of affect, for the capitalized "Opera" recalls the famous Paris Opera Ballet, the world's oldest ballet school and, in Dickinson's time, a key source of skilled dancers for the American stage (see Martin, "Naked Females").

66. Many have discussed Susan's contributions to the poems now attributed to Emily Dickinson. Regarding her comments on this poem particularly, see Martha Nell Smith, *Rowing in Eden: Rereading Emily Dickinson* (Austin: University of Texas Press, 1992).

67. Smith, *Rowing in Eden*, 181.

68. Franklin, *The Poems of Emily Dickinson*, 161.

69. This variant, sent to T. W. Higginson in 1862, is the latest available. The poet had earlier revised "Sleep" in line 4 to "Lie," but here she reverts it. The earlier version copied into Fascicle 6 also places a comma and line break after "satin."

70. Anne-Lise François, *Open Secrets: The Literature of Uncounted Experience* (Stanford, Calif.: Stanford University Press, 2007), 204.

71. Wallace Stevens, "The Snow Man," in *The Collected Poems of Wallace Stevens* (New York: Random House, 1990), 10.

72. Martha Nell Smith and Lara Vetter, "Emily Dickinson Writing a Poem," *Dickinson Electronic Archives*, 1999, archive.emilydickinson.org/safe/index.html.

73. Cf. Smith, *Rowing in Eden*, 181.

74. Other miniature editions include Franklin's book-bound facsimile of the "Master" letters, Smith and Hart's *Open Me Carefully*, Werner's *Radical Scatters* and *The Gorgeous Nothings*.

75. Smith and Vetter's transcription omits the "I" before "always," but it is plainly evident in the holograph image. (Smith and Vetter, "Emily Dickinson Writing a Poem").

76. Thomas Wentworth Higginson, "An Open Portfolio," *Christian Union* 42, no. 13 (September 25, 1890): 393.

77. Susan and Emily Dickinson appear to have shared a view that good poetry gives a chill no fire can assuage. Emily writes in a letter to Higginson, "If I read a book and it makes my whole body so cold no fire can warm me I know *that* is poetry" (*L* 342a).

78. Mabel Loomis Todd, "Preface," in *Poems by Emily Dickinson: Second Series,* ed. Mabel Loomis Todd and T. W. Higginson (Boston: Roberts Brothers, 1891), 5–6.

79. Theodora Van Wagenen Ward, "Characteristics of the Handwriting," in *The Poems of Emily Dickinson,* ed. Thomas Johnson (Cambridge, Mass.: Harvard University Press, 1955), liv, lv.

80. Todd, "Preface," 6.

81. Mabel Loomis Todd, "Introductory," in *Letters of Emily Dickinson,* ed. Mabel Loomis Todd (Boston: Roberts Brothers, 1894), viii.

82. Todd, *Letters of Emily Dickinson,* 157–59.

83. Roland Barthes, *Camera Lucida: Reflections on Photography,* trans. Richard Howard (New York: Hill and Wang, 1982), 79.

84. Ward, "Characteristics of the Handwriting," lix.

85. Barthes, *Camera Lucida,* 14.

86. Other contenders include Hopkins's stress mark, Lowell's ellipsis, and Ammons's colon. Whether these punctuations become so strongly identified with their poets as Dickinson's dash has been will depend as much upon the future reputations of the poets as on the specific reading habits that develop around them.

87. Smith, *Rowing in Eden,* 19, 110.

88. Paul Crumbley, *Inflections of the Pen: Dash and Voice in Emily Dickinson* (Lexington: University Press of Kentucky, 1997), 1–2.

89. See Stephen Toumlin and June Goodfield, *The Architecture of Matter* (Chicago: University of Chicago Press, 1962), 307–37.

90. *Open Me Carefully: Emily Dickinson's Intimate Letters to Susan Huntington Dickinson,* ed. Martha Nell Smith and Ellen Louise Hart (New York: Paris Press, 1998), xvii.

91. Martha Nell Smith, "Omissions Are Not Accidents: Erasures & Cancellations in Emily Dickinson's Manuscripts," introduction to "Mutilations: What Was Erased, Inked Over, and Cut Away," ed. Martha Nell Smith and Jarome McDonald, *Dickinson Electronic Archives,* 2000, archive.emilydickinson.org/mutilation/mintro.html.

92. Smith, "Omissions Are Not Accidents."

93. Smith, "Omissions Are Not Accidents."

94. Eve Kosofsky Sedgwick, *Epistemology of the Closet* (Berkeley: University of California Press, 1990), 12.

95. Jackson, *Dickinson's Misery,* 136.

96. Jackson, *Dickinson's Misery,* 135.

97. Jonathan Morse, "Conduct Book and Serf: Emily Dickinson Writes a Word," *The Emily Dickinson Journal* 16, no. 1 (2007): 58.

98. Domhnall Mitchell, *Emily Dickinson: Monarch of Perception* (Amherst: University of Massachusetts Press, 2000), 172.

99. Ibid.

100. Janet Holmes, *The Ms of My Kin* (Exeter, UK: Shearsman, 2009), 24.

101. Holmes, *The Ms of My Kin*, 169.

102. Faith Barrett, *To Fight Aloud Is Very Brave: American Poetry and the Civil War* (Amherst: University of Massachusetts Press, 2012), 290.

103. John Paul Ricco, "Name No One Man," *Parallax* 11, no. 2 (2005): 97.

104. Michael Magee, response to Charles Bernstein in "The Flarf Files," *Electronic Poetry Center,* August 2003, writing.upenn.edu/epc/authors/bernstein/syllabi/readings/flarf.html.

105. Holmes, *The Ms of My Kin*, 98–99. The proximal reference here is to the U.S. government's presentation of flimsy evidence that Saddam Hussein had weapons of mass destruction in order to justify the 2003 invasion of Iraq.

106. Rick Snyder, "The New Pandemonium: A Brief Overview of Flarf," *Jacket* 31 (October 2006), http://jacketmagazine.com/31/snyder-flarf.html.

107. Snyder, "The New Pandemonium."

108. Michael Magee, *My Angie Dickinson* (La Laguna, Spain: Zasterle, 2006), 36.

109. Jennifer Ashton, "Sincerity and the Second Person: Lyric after Language Poetry," *Interval(le)s* 2, no. 2–3, no. 1 (Fall 2008–Winter 2009), 99, http://labos.ulg.ac.be/cipa/wp-content/uploads/sites/22/2015/07/8_ashton.pdf.

110. Magee, *My Angie Dickinson*, 4.

111. Angie Dickinson worked with a number of important directors in the 1960s, making innovative films just after the Production Code was abolished. These included *The Killers* (Don Siegel, 1964), *The Chase* (Arthur Penn, 1966), and *Point Blank* (John Boorman, 1967). During the 1970s, however, she was increasingly cast as an exploitation element in B-movies such as *Pretty Maids All in a Row* (Roger Vadim, 1971) and *Big Bad Mama* (Steve Carver, 1974). By the time Brian de Palma cast her in *Dressed to Kill* (1980), she was better known as a sex object than for her considerable acting abilities.

112. Gary Sullivan, "Jacket Flarf Feature: Introduction," *Jacket* 30, July 2006, http://jacketmagazine.com/30/fl-intro.html.

2. CHANCE

1. RAND Corporation, "Foreword to the Online Edition," in *A Million Random Digits with 100,000 Normal Deviates,* online version, https://www.rand.org/pubs/monograph_reports/MR1418/index2.html.

2. In a lecture on "The Poetics of Chance" and elsewhere, Mac Low discusses the relations among chance operations, his anarcho–pacifist politics, and his Buddhist/Taoist beliefs; see Jackson Mac Low, "The Poetics of Chance and the Politics of Simultaneous Spontaneity, or the Sacred Heart of Jesus (Revised & Abridged)," July 12,

1975, in *Talking Poetics from the Naropa Institute*, ed. Anne Waldman and Marilyn Webb (Boulder, Colo,: Shambhala, 1978), 171–94. Hereafter, this work will be cited in the text as "Poetics."

3. Wherever possible, I wish to maintain clear distinctions among this chapter's key terms—chance, randomness, contingency, arbitrariness—but their shared opposition to determinacy and rational systems of causation often makes them difficult to distinguish from one another. Philosophers and literary critics have not reached consensus about the subtle differences among them. Like other vocabularies of lyric exemption, their specific meanings matter less to poets than their collective opposition to some idea of rationalism. For a preliminary anatomy of their differences in experimental poetics, see Brian McHale, "Poetry as Prosthesis," *Poetics Today* 21, no. 1 (Spring 2000): 1–32.

4. Among others, Mac Low's mentor John Cage understood chance operations as a detachment from history. For Cage, "history is the story of original actions." So, when a composer deploys sounds "not intended" by personal choice, the composition is "uninformed about history." See John Cage, *Silence: Lectures and Writings* (Middletown, Conn.: Wesleyan University Press, 2011; originally 1961), 74, 14. Ellen Zweig argues that, for Mac Low, chance operations "take us away from our Western preoccupation with causality and time sequence" and thus detach the text from "our culture and our personal history" ("Jackson Mac Low: The Limits of Formalism," *Poetics Today* 3, no. 3 [1982]: 85, 83). Mac Low himself links chance operations with a Buddhist metaphysics. He thus sees "*some* kind of cause" influencing his chance operations, "but it's different from the time-linear cause going from past to future" ("Poetics," 176).

5. Jackson Mac Low, "Stein 100: A Feather Likeness of the Justice Chair," *poets.org*, The Academy of American Poets, September 20, 1999, poets.org/poetsorg/poem/stein-100-feather-likeness-justice-chair.

6. Gertrude Stein, *Tender Buttons: The Corrected Centennial Edition*, ed. Seth Perlow (San Francisco: City Lights, 2014), 20. Hereafter, this work will be cited in the text as *TB*. Mac Low's note to "Stein 100" misquotes this paragraph's sixth word as "any." In fact, the text reads "the," and Mac Low's poem itself indicates that the procedure used the correct word.

7. Christopher Funkhouser, *Prehistoric Digital Poetry: An Archaeology of Forms, 1959–1995*, Modern and Contemporary Poetics (Tuscaloosa: University of Alabama Press, 2007), 80.

8. Lisa Siraganian, "Out of Air: Theorizing the Art Object in Gertrude Stein and Wyndham Lewis," *Modernism/modernity* 10, no. 4 (2003): 663. In "Poetry and Grammar," Stein reserves her praise for the sparest punctuation marks, the period and comma, calling them interesting only when they "commence breaking up things in arbitrary ways" ("Poetry and Grammar," in *Writings, 1932–1946* [New York: Library of America, 1998], 318).

9. Louis Cabri, "'Rebus Effort Remove Government': Jackson Mac Low, *Why?/ Resistance*, Anarcho/Pacifism," *Crayon* 1, no. 1 [special issue: *Festschrift for Jackson Mac Low's 75th Birthday*, ed. Andrew Levy and Bob Harrison] (1997): 58.

10. Funkhouser, *Prehistoric Digital Poetry*, 79.

11. Jackson Mac Low, "Museletter," in *The L=A=N=G=U=A=G=E Book*, ed. Bruce Andrews and Charles Bernstein (Carbondale: Southern Illinois University Press, 1984), 26.

12. Charles Bernstein, *Content's Dream: Essays 1979–1984* (Evanston, Ill.: Northwestern University Press, 2001; originally 1986), 252.

13. See Jonathan Culler, *Theory of the Lyric* (Cambridge, Mass.: Harvard University Press, 2015), 77–85.

14. Theodor W. Adorno, "On Lyric Poetry and Society," in *Notes to Literature*, vol. 1, ed. Rolf Tiedemann, trans. Shierry Weber Nicholsen (New York: Columbia University Press, 1991), 38, 39.

15. Theodor W. Adorno, *Aesthetic Theory*, ed. and trans. Robert Hullot-Kentor, Theory and History of Literature 88 (Minneapolis: University of Minnesota Press, 1997), 221.

16. Peter Middleton, *Physics Envy: American Poetry and Science in the Cold War and After* (Chicago: University of Chicago Press, 2015), 208–9.

17. I am grateful to Charles O. Hartman for his emails about DIASTEX5 and to Ron Starr for making an online engine for diastic reading, along with a description of the process, which have now disappeared. Hartman discusses his programs for computer poetry in *Virtual Muse: Experiments in Computer Poetry* (Middletown, Conn.: Wesleyan University Press, 1996). A modified "engine" for diastic reading can be found at http://www.eddeaddad.net/ediastic/.

18. Jackson Mac Low, *Asymmetries 1–260: The First Section of a Series of 501 Performance Poems* (New York: Printed Editions, 1980), 246, 248, 249.

19. Jacques Derrida, "My Chances / *Mes Chances*: A Rendezvous with Some Epicurean Stereophonies," in *Taking Chances: Derrida, Psychoanalysis, and Literature*, ed. Joseph H. Smith and William Kerrigan (Baltimore, Md.: Johns Hopkins University Press, 1984), 25.

20. Derrida, "My Chances," 2. (emphasis original).

21. Jackson Mac Low, *Thing of Beauty: New and Selected Works*, ed. Anne Tardos (Berkeley: University of California Press, 2008), 376.

22. Bernstein, *Content's Dream*, 235.

23. Mac Low, *Asymmetries 1–260*, 245.

24. Mac Low, "Stein 100," line 73.

25. Jackson Mac Low, "Pleasant to Be Repeating Very Little of This (Stein 32)," in *Thing of Beauty*, 389.

26. Mac Low, "Stein 100," line 18.

27. Tyrus Miller, "'Who am I?' or, The Reader as Anarchist: (On Jackson Mac Low's *Stanzas for Iris Lezak*)," *American Letters & Commentary* 8 (1996): 43.

28. Steven Meyer, *Irresistible Dictation: Gertrude Stein and the Correlations of Writing and Science* (Stanford, Calif.: Stanford University Press, 2001), 122.

29. Jennifer Ashton, *From Modernism to Postmodernism: American Poetry and Theory in the Twentieth Century* (Cambridge: Cambridge University Press, 2005), 28.

30. Phoebe Stein Davis, "'Even Cake Gets to Have Another Meaning': History, Narrative, and 'Daily Living' in Gertrude Stein's World War II Writings," *Modern Fiction Studies* 44, no. 3 (1998): 574.

31. Kelly Wagers, "Gertrude Stein's 'Historical Living,'" *Journal of Modern Literature* 31, no. 3 (2008): 23, 25.

32. Gertrude Stein, "Reflection on the Atomic Bomb," *Yale Poetry Review* 7 (December 1947): 3–4.

33. For a reading of information overload in Stein's work, including her "Reflection on the Atomic Bomb," see Paul Stephens, *The Poetics of Information Overload: From Gertrude Stein to Conceptual Writing* (Minneapolis: University of Minnesota Press, 2015).

34. See Gertrude Stein, "Composition as Explanation" (1925), in *What Are Masterpieces*, ed. Robert Haas (New York: Pitman, 1970), 25–38.

35. Kristin Bergen, "'Dogs Bark': War, Narrative, and Historical Syncopation in Gertrude Stein's Late Work," *Criticism* 57, no. 4 (Fall 2015): 615, 615, 613.

36. Seth Perlow, "A Note on the Text," in Stein, *Tender Buttons: The Corrected Centennial Edition*, 95.

37. Roland Barthes, *Camera Lucida: Reflections on Photography*, trans. Richard Howard (New York: Hill and Wang, 1982), 6. Barthes associates this disorder not only with photography but also with the arbitrary forms of language, which Stein's works underscore.

38. Juliana Spahr, "Afterword," in Stein, *Tender Buttons: The Corrected Centennial Edition*, 110.

39. Adorno, *Aesthetic Theory*, 156.

40. Adorno, *Aesthetic Theory*, 98.

41. Adorno, *Aesthetic Theory*, 98.

42. Adorno, *Aesthetic Theory*, 156.

43. Gertrude Stein, *Tender Buttons: Objects, Food, Rooms* (New York: Claire Marie, 1914), 16.

44. The most compelling theories of authorial intention are also the most extreme, but chance poses the same difficulty for both sides of the controversy. Whether you share the New Critics' strict textualism or believe a text only ever means what its author intends, the idea that it might mean something by chance must be disavowed, not to mention the possibility that chance provides a basis for meaning. This shared disavowal might seem innocuous if not for the long history of literary experiments with chance. For representative statements from the two poles of the intentionality debate, see: William K. Wimsatt and Monroe C. Beardsley, "The Intentional Fallacy,"

Sewanee Review 54 (1946): 468–88; and Steven Knapp and Walter Benn Michaels, "Against Theory," *Critical Inquiry* 8, no. 4 (Summer 1982): 723–42.

45. Marjorie Perloff, *The Poetics of Indeterminacy: Rimbaud to Cage* (Chicago: Northwestern University Press, 1983), 100.

46. Spahr, "Afterword," 112–13. For her part, Spahr views the "bourgeois interior" of *Tender Buttons* not only as reflecting Stein's interest in domestic life but also as "registering, whether intended or not, the systemic crisis that was imperial globalization" in early twentieth-century Europe (110, 119).

47. Perloff, *The Poetics of Indeterminacy*, 101, 102.

48. Sianne Ngai, "The Cuteness of the Avant-Garde," *Critical Inquiry* 31, no. 4 (Summer 2005): 815.

49. Ngai, "The Cuteness of the Avant-Garde," 811, 816, 829.

50. Ngai, "The Cuteness of the Avant-Garde," 822.

51. Perloff, *The Poetics of Indeterminacy*, 99, 102.

52. Barbara Barracks, "Passing the Word Along: Mac Low in Brief," *L=A=N=G=U=A=G=E* 1, no. 2 (April 1978): 24–25.

53. RAND Corporation, *A Million Random Digits with 100,000 Normal Deviates* (Glencoe, Ill.: Free Press, 1955), ix.

54. Nicholas Metropolis, "The Beginning of the Monte Carlo Method," *Los Alamos Science* 15 (1987): 127.

55. Metropolis, "The Beginning of the Monte Carlo Method," 127.

56. George W. Brown, *History of RAND's Random Digits—Summary*, RAND paper P-113, June 1949, 1, rand.org/content/dam/rand/pubs/papers/2008/P113.pdf.

57. Willis H. Ware, *RAND and the Information Evolution* (Santa Monica, Calif.: RAND Corporation, 2008), 90.

58. Ware, *RAND and the Information Evolution*, 89.

59. Alex Wellerstein, *Oppenheimer: A Life, April 27, 1904–February 18, 1967, an Online Centennial Exhibit of J. Robert Oppenheimer*, University of California, Berkeley, 2004, beck2.med.harvard.edu/week12/oppenheimer.pdf.

60. Trade presses have published some useful, if adulatory, histories of these mid-century technological research labs: John Gertner, *The Idea Factory: Bell Labs and the Great Age of American Innovation* (New York: Penguin, 2013); Michael A. Hiltzik, *Dealers of Lightning: Xerox PARC and the Dawn of the Computer Age* (New York: Harper Business, 2000); and Sharon Weinberger, *The Imagineers of War: The Untold Story of DARPA, the Pentagon Agency That Changed the World* (New York: Knopf, 2017).

61. Claude E. Shannon, "A Mathematical Theory of Communication," *The Bell System Technical Journal* 27 (1948): 379–423 and 623–56, at 379.

62. Michel Serres has developed a philosophical account of noise as a necessary backdrop of communication (*Hermes: Literature, Science, Philosophy*, ed. Josué V. Harari and David F. Bell [Baltimore, Md.: Johns Hopkins University Press, 1982]).

63. Christophe Andrieu et al., "An Introduction to MCMC for Machine Learning," *Machine Learning* 50, no. 1/2 (2003): 5–43.

64. Bernard Joseph Carr, *The Basic Physics of an Atomic Bomb*, produced by Cascade Pictures of California for the Armed Special Forces Weapons Project (1950; Project Number 5001, sanitized Department of Energy Version ID 0800091), YouTube video, 20:29, youtube.com/watch?v=0WdJZGw4wp8.

65. Derrida, "My Chances," 6.

66. The link between computers and military activity is easier to recognize but still occasions infelicitous analogies. In *Computer Poems* (1973), for instance, Richard Bailey claims that "Computer poetry is warfare carried out by other means" (quoted in Funkhouser, *Prehistoric Digital Poetry,* 79). One wonders how Bailey might respond to poetry that involves actual military equipment. See Martin van Creveld, *Wargames: From Gladiators to Gigabytes* (New York: Cambridge University Press, 2013).

67. Mac Low too finds his sense of history reshaped by uncertainties about the bomb. Reflecting on the 1960s, he notes that "the threat of atomic war had hung over the world for a decade and a half" (*Asymmetries 1–260,* 241).

68. Jackson Mac Low, "Who Is Showing Us What Happened in This Corner? (Stein 103)," *Deluxe Rubber Chicken* no. 7 (May 2001), *Electronic Poetry Center,* writing.upenn.edu/epc/ezines/deluxe/seven/103.html.

69. Discussing a 1947 political essay by Mac Low, Louis Cabri links the question mark with political anxieties: because "Mac Low does not know what direct action to recommend," he instead "reawakens the suppressed question mark—hence panic—in the title phrase of Lenin's *What Is To Be Done,* the essay that was intended to provide the authoritative answer, without question, for revolution" ("Rebus Effort," 55).

70. William Faulkner, "Banquet Speech," Nobel Banquet, City Hall, Stockholm, Sweden, December 10, 1950, *Nobelprize.org: The Official Website of the Nobel Prize,* nobelprize.org/nobel_prizes/literature/laureates/1949/faulkner-speech.html.

71. W. H. Auden, *The Age of Anxiety: A Baroque Eclogue,* ed. Alan Jacobs (Princeton, N.J.: Princeton University Press, 2011; originally 1947), 3.

72. On Stein and automatic or unintentional writing, see B. F. Skinner's famous essay "Has Gertrude Stein a Secret?" (1934), in *Critical Essays on Gertrude Stein,* ed. Michael J. Hoffman (Boston: G. K. Hall, 1986), 64–70.

73. Gertrude Stein, "An American and France," in *What Are Masterpieces,* 59.

74. Walt Whitman, "Song of Myself," in *The Complete Poems* (New York: Penguin, 2005), 63.

75. Mark Noble, "Whitman's Atom and the Crisis of Materiality in the Early *Leaves of Grass,*" *American Literature* 81.2 (June 2009): 253.

76. Mina Loy, "Gertrude Stein," *Transatlantic Review* 2 (1924): 305.

77. Gertrude Stein, *Everybody's Autobiography* (New York: Exact Change, 2004; originally 1937), 22.

78. Allen Ginsberg, "Howl," in *Howl and Other Poems* (San Francisco: City Lights, 1956), 11.

79. John Berryman, *The Dream Songs* (New York: Farrar, Straus and Giroux, 1959), 55.

80. In *Toy Medium*, Daniel Tiffany produces a genealogy of poems about nuclear weapons but does not link this theme to national identity; he cites work by Ginsberg, Auden, Kinnell, Corso, Dickey, and others. Daniel Tiffany, *Toy Medium: Materialism and Modern Lyric* (Berkeley: University of California Press, 2000), 217–44.

81. Lucretius (Titus Lucretius Carus), *De rerum natura* 222–24, in *On the Nature of Things*, trans. Robert Allison (London: Arthur L. Humphreys, 1919).

82. Lucretius, *De rerum natura* 289–90.

83. The most consequential modern reading of the *clinamen* as a basis for free will appears in Karl Marx's doctoral dissertation, "The Difference between Democritean and Epicurean Philosophy of Nature" (1841), in *Karl Marx, Frederick Engels: Collected Works,* vol. 1, *Marx: 1835–1843,* trans. Richard Dixon (New York: International, 1975), 25–107. This work proved formative for the accommodation of human freedom in Marx's later materialism. The most influential modern reading of Lucretian philosophy in general is Michel Serres, *The Birth of Physics,* trans. Jack Hawkes (Manchester, UK: Clinamen Press, 2001; originally 1977).

84. Alison James, "Aleatory Poetics," in *The Princeton Encyclopedia of Poetry and Poetics,* 4th ed., ed. Roland Greene et al. (Princeton, N.J.: Princeton University Press, 2012), 33.

85. Jackson Mac Low, "Reading a Selection from *Tender Buttons,*" L=A=N=G=U=A=G=E 1, no. 16 (December 1978): 9.

86. Bernstein, *Content's Dream,* 252.

87. Steve McCaffery, *The Darkness of the Present: Poetics, Anachronism, and the Anomaly* (Tuscaloosa: University of Alabama Press, 2012), 48, 50.

88. Jed Rasula, "Statement on Reading in Writing," in Andrews and Bernstein, *The L=A=N=G=U=A=G=E Book,* 53.

89. Ron Silliman, "IF BY 'WRITING' WE MEAN LITERATURE (if by 'literature' we mean poetry (if . . .)) . . . ," in Andrews and Bernstein, *The L=A=N=G=U=A=G=E Book,* 167.

90. Ron Silliman, "Disappearance of the Word, Appearance of the World," in Andrews and Bernstein, *The L=A=N=G=U=A=G=E Book,* 127.

91. Derrida, "My Chances," 2.

92. Adorno, *Aesthetic Theory,* 108, 115.

93. Allen Grossman, "Summa Lyrica: A Primer of the Commonplaces in Speculative Poetics," in *The Sighted Singer: Two Works on Poetry for Readers and Writers* (Baltimore, Md.: Johns Hopkins University Press, 1992), 224.

94. Jennifer Scappettone, "'Più mOndo i: / tUtti!': Traffics of Historicism in Jackson Mac Low's Contemporary Lyricism," *Modern Philology* 105, no. 1 (2007): 188.

95. McCaffery, *The Darkness of the Present,* 47, 43, 44.

96. Scappettone, "Traffics of Historicism," 205, 200.

97. Jackson Mac Low, "Some Ways Philosophy Has Helped to Shape My Work" (1983), in *A Guide to Poetics Journal: Writing in the Expanded Field, 1982–1998,* ed. Lyn Hejinian and Barrett Watten (Middletown, Conn.: Wesleyan University Press, 2013), 121, 123.

98. Jackson Mac Low, "Response to Piombino," June 1997, *Electronic Poetry Center,* writing.upenn.edu/epc/authors/maclow/piombino.html.

99. Jackson Mac Low, "Interview with Charles Bernstein," *The LINEbreak Program,* 1995, *Electronic Poetry Center* and *PennSound,* 29:00, media.sas.upenn.edu/ pennsound/groups/LINEbreak/Mac-Low/Mac-Low-Jackson_LINEbreak_NY_ 1995.mp3.

100. Mac Low, "Some Ways Philosophy Has Helped," 122.

101. Mac Low described himself as an anarchist, a pacifist, and a left libertarian, noting that "Aristotle's view of history . . . helped immunize me against political arguments based on dialectical or historical materialism" ("Some Ways Philosophy Has Helped," 121). More overtly political theories of chance stem from Louis Althusser's discussions of aleatory materialism, especially "The Underground Current of the Materialism of the Encounter," in *Philosophy of the Encounter: Later Writings, 1978–1987,* trans. G. M. Goshgarian (New York: Verso, 2006), 163–207. See also Warren Montag, *Louis Althusser* (New York: Palgrave Macmillan, 2003), and Antonio Negri, "Notes on the Evolution of the Thought of the Later Althusser," trans. Olga Vasile, in *Postmodern Materialism and the Future of Marxist Theory: Essays in the Althusserian Tradition,* ed. Antonio Callari and David F. Ruccio (Middletown, Conn.: Wesleyan University Press, 1996), 51–68.

102. Andrew Levy and Jackson Mac Low, "PhillyTalks 2: Poetry Reading and Discussion," November 5, 1997, University of Pennsylvania, *PennSound,* media.sas.upenn .edu/pennsound/groups/phillytalks/02/PhillyTalks2_Complete-Recording_11 -05-97_UPenn.mp3.

103. Scappettone, "Traffics of Historicism," 192.

104. Mac Low, "Interview with Charles Bernstein."

105. Steve McCaffery expresses surprising credulity about the redemptive alchemy of Mac Low's technique. For him, "it brings up the possibility for numerous strategic readings. . . . A survivor of the Shoah for example might choose 'Auschwitz' as a theme word and trace it through Hitler's *Mein Kampf*. . . . A feminist might take 'Molly Bloom' and read through *Ulysses*" (*The Darkness of the Present,* 218n4). I linger with the difficulty of reading Mac Low's texts and the unhappy origins of his equipment in order to maintain some skepticism about the reparative goals of his work.

106. Cabri, "Rebus Effort," 61.

107. On Stein and feminism, see: Ellen Berry, *Curved Thought and Textual Wandering: Gertrude Stein's Postmodernism* (Ann Arbor: University of Michigan Press, 1992); Harriet Chessman, *The Public Is Invited to Dance: Representation, the Body, and Dialogue in Gertrude Stein* (Stanford, Calif.: Stanford University Press, 1989); Marianne DeKoven, *A Different Language: Gertrude Stein's Experimental Writing* (Madison:

University of Wisconsin Press, 1983); and Catharine R. Stimpson, "The Mind, the Body, and Gertrude Stein," *Critical Inquiry* 3, no. 3 (1977): 489–506. On Stein and lesbianism, see: Karin Cope, *Passionate Collaborations: Learning to Live with Gertrude Stein* (Victoria, B.C.: ELS Editions, 2005); Kathryn R. Kent, *Making Girls into Women: American Women's Writing and the Rise of Lesbian Identity* (Durham, N.C.: Duke University Press, 2002); and Catharine R. Stimpson, "Gertrude Stein and the Lesbian Lie," in *American Women's Autobiography: Fea(s)ts of Memory*, ed. Margo Culley (Madison: University of Wisconsin Press, 1992), and "Gertrude Stein and the Transposition of Gender," in *The Poetics of Gender*, ed. Nancy K. Miller (New York: Columbia University Press, 1986). On Stein and domesticity, see: Sara Blair, "Home Truths: Gertrude Stein, 27 Rue de Fleurus, and the Place of the Avant-Garde," *American Literary History* 12, no. 3 (2000): 417–33; Phoebe Stein Davis, "'Even Cake Gets to Have Another Meaning'"; and Steven Watson, *Prepare for Saints: Gertrude Stein, Virgil Thomson, and the Mainstreaming of American Modernism* (Berkeley: University of California Press, 1995). On Stein's Jewishness, see Maria Damon, "Gertrude Stein's Jewishness, Jewish Social Scientists, and the 'Jewish Question,'" *Modern Fiction Studies* 42 (1996): 489–506. On her Vichy relationships and Zionism, see: Wanda Van Dusen, "Portrait of a National Fetish: Gertrude Stein's 'Introduction to the Speeches of Maréchal Pétain,'" *Modernism/modernity* 3, no. 1 (1996): 69–92; and Barbara Will, *Unlikely Collaboration: Gertrude Stein, Bernard Faÿ, and the Vichy Dilemma* (New York: Columbia University Press, 2011).

108. Mac Low, "Some Ways Philosophy Has Helped," 121.

109. Bernstein, *Content's Dream*, 256.

110. Bernstein, *Content's Dream*, 253.

111. Brett Bourbon, "What Is a Poem?" *Modern Philology* 105, no. 1 (2007): 27–28.

112. Altieri understands "most of the innovative work of the [1950s] as elaborating a 'logic of contingency,' whose parameters are still being worked out by contemporary poets and critics." However, his readings of O'Hara, Creeley, and Plath emphasize the poet's personhood more faithfully than the *Stein* poems warrant (Charles Altieri, "Contingency as Compositional Principle in Fifties Poetics," in *The Scene of My Selves: New Work on New York School Poets*, ed. Terence Diggory and Stephen Paul Miller [Orono, Maine: National Poetry Foundation, 2001], 359).

113. Some critiques of this privilege of raw data appear in Lisa Gitelman, ed., *"Raw Data" Is an Oxymoron* (Cambridge, Mass.: The MIT Press, 2013).

3. ANONYMITY

1. Larry Rivers, "Speech Read at Frank O'Hara's Funeral," in *Homage to Frank O'Hara*, ed. Bill Berkson and Joe LeSueur (Bolinas, Calif.: Big Sky, 1988), 138.

2. In the first major study of the poet, *Frank O'Hara: Poet among Painters* (Chicago: University of Chicago Press, 1977), Marjorie Perloff acknowledges his reputation as a "minor artist, memorable less for his actual achievement than for his colorful life and his influence on others" (2). Perloff's book helped canonize his poetry, but as

her title suggests, she reads him in terms of his sociality. More recently, Lytle Shaw's *Frank O'Hara: The Poetics of Coterie* (Iowa City: University of Iowa Press, 2006) and Andrew Epstein's *Beautiful Enemies: Friendship in Postwar American Poetry* (New York: Oxford University Press, 2006) likewise discuss O'Hara in terms of his many friendships, as do several of the essays in *Frank O'Hara Now: New Essays on the New York Poet*, ed. Robert Hampson and Will Montgomery (Liverpool, UK: Liverpool University Press, 2010).

3. Keston Sutherland, "Close Writing," in Hampson and Montgomery, *Frank O'Hara Now*, 122, 129.

4. Frank O'Hara, *The Collected Poems of Frank O'Hara*, ed. Donald Allen (New York: Alfred A. Knopf, 1971), 32. Hereafter, this work will be cited in the text as *Collected*.

5. Among others, see Sherry Turkle, *Alone Together: Why We Expect More from Technology and Less from Each Other* (New York: Basic Books, 2011).

6. Andrew Epstein associates O'Hara with a "reorientation of poetry to the everyday" that has now become common among American poets: poetry brings critical attention to quotidian objects and experiences that normally get ignored, including telephones and telephone calls (*Attention Equals Life: The Pursuit of the Everyday in Contemporary Poetry and Culture* [New York: Oxford University Press, 2016], 2).

7. For a full account of O'Hara's gossip, see Chad Bennett, *Word of Mouth: Gossip and American Poetry* (Baltimore, Md.: Johns Hopkins University Press, 2018).

8. Perloff, *Frank O'Hara*, 29.

9. Redell Olsen, "Kites and Poses: Attitudinal Interfaces in Frank O'Hara and Grace Hartigan," in Hampson and Montgomery, *Frank O'Hara Now*, 189.

10. Perloff, *Frank O'Hara*, 26.

11. Rod Mengham, "French Frank," in Hampson and Montgomery, *Frank O'Hara Now*, 56; David Herd, "Stepping Out with Frank O'Hara," in Hampson and Montgomery, *Frank O'Hara Now*, 70.

12. Hazel Smith, *Hyperscapes in the Poetry of Frank O'Hara: Difference, Homosexuality, Topography* (Liverpool, UK: Liverpool University Press, 2000), 143.

13. Oren Izenberg, *Being Numerous: Poetry and the Ground of Social Life* (Princeton, N.J.: Princeton University Press, 2011), 136.

14. Izenberg, *Being Numerous*, 137.

15. Edward Lucie-Smith, "An Interview with Frank O'Hara," October 1965, in *Frank O'Hara: Standing Still and Walking in New York*, ed. Donald Allen (San Francisco: Grey Fox, 1983), 26.

16. Alfred Leslie, *USA Poetry: Frank O'Hara*, directed by Richard O. Moore, March 5, 1966, YouTube video, 15:56, https://www.youtube.com/watch?v=344TyqLlSFA, available in the DVD collection *Cool Man in a Golden Age: Selected Films, 1959–2007* (London: Lux, 2009).

17. Discussing the film, John Ashbery identifies the caller as Brodey ("Interview with Ron Padgett," *Oral History Initiative: On Frank O'Hara*, Woodberry Poetry

Room, Harvard University, Cambridge, Mass., April 5, 2011, YouTube video, 1:17:43, youtube.com/watch?v=0acw2wX5nac).

18. Michael Magee, *Emancipating Pragmatism: Emerson, Jazz, and Experimental Writing* (Tuscaloosa: University of Alabama Press, 2004), 143.

19. Sutherland, "Close Writing," 122.

20. Izenberg, *Being Numerous*, 136.

21. Joe Brainard, "Frank O'Hara," in Berkson and LeSeur, *Homage to Frank O'Hara*, 168.

22. Those who would confine "apostrophe" to dead or nonhuman addressees forget that, in Quintilian's rhetoric, from which the term enters English, it refers to any address not directed at the judges of an oration, including others attending in person. I nonetheless use the more capacious "address" unless more specificity is warranted. Except where a poem addresses the reader directly, this triangulated structure characterizes all poetic address. See Jonathan Culler, *Theory of the Lyric* (Cambridge, Mass.: Harvard University Press, 2015), 186.

23. Jonathan Culler, *The Pursuit of Signs: Semiotics, Literature, Deconstruction* (Ithaca, N.Y.: Cornell University Press, 1981), 137.

24. Ann Keniston, *Overheard Voices: Address and Subjectivity in Postmodern American Poetry* (New York: Routledge, 2006), 4.

25. Culler, *Theory of the Lyric*, 243.

26. William Carlos Williams, "This Is Just to Say," in *Selected Poems* (New York: New Directions, 1985), 74.

27. Bill Berkson, "Re: as planned . . . ," email to the author, August 6, 2011.

28. Ashbery, "Interview with Ron Padgett."

29. Keniston, *Overheard Voices*, 8.

30. Allen Ginsberg, "Howl," in *Howl and Other Poems* (San Francisco: City Lights, 1956), 24.

31. Allen, "Editor's Note," in *Collected*, v.

32. Allen provides no note to this poem in *Collected* (an apparent oversight), but given its placement in the collection, it was likely written in 1952.

33. Allen notes that, two years after this poem was written, Miller would publish O'Hara's *Odes* "with silk-screen prints by Mike Goldberg" (*Collected*, 542).

34. Sutherland, "Close Writing," 128.

35. Allen Grossman, "Summa Lyrica: A Primer of the Commonplaces in Speculative Poetics," in *The Sighted Singer: Two Works on Poetry for Readers and Writers* (Baltimore, Md.: Johns Hopkins University Press, 1992), 239.

36. Lucie-Smith, "Interview," 18.

37. Michel Foucault, *Discipline and Punish: The Birth of the Prison*, trans. Alan Sheridan (New York: Random House, 1977; originally 1975), 205.

38. On O'Hara's impersonal personhood, especially in relation to his consumerism, see Michael Clune, "'Everything We Want': Frank O'Hara and the Aesthetics of Free Choice," *PMLA*, no. 1201 (January 2005): 181–96.

39. Michael Warner, *The Trouble with Normal: Sex, Politics, and the Ethics of Queer Life* (Cambridge, Mass.: Harvard University Press, 2000), 166.

40. Grossman, "Summa Lyrica," 212.

41. Exceptions include: Johanna Drucker, *Graphesis: Visual Forms of Knowledge Production* (Cambridge, Mass.: Harvard University Press, 2014); Lisa Gitelman, *Paper Knowledge: Toward a Media History of Documents* (Durham, N.C.: Duke University Press, 2014); and Matthew G. Kirschenbaum, *Track Changes: A Literary History of Word Processing* (Cambridge, Mass.: Harvard University Press, 2016).

42. John Stuart Mill, "Thoughts on Poetry and Its Varieties," *CUWS: Classical Utilitarianism Web Site*, University of Texas, laits.utexas.edu/poltheory/jsmill/diss-disc/poetry/poetry.s01.html.

43. Mutlu Blasing defines the lyric in terms of this solipsistic address, "where an 'I' talks to itself or to nobody in particular and is not primarily concerned with narrating a story or dramatizing an action" (*Lyric Poetry: The Pain and the Pleasure of Words* [Princeton, N.J.: Princeton University Press, 2007], 2). Another notable conflation of apostrophe and soliloquy as signs of the lyric appears in Northrop Frye, *Anatomy of Criticism: Four Essays* (Princeton, N.J.: Princeton University Press, 1957), 246–51.

44. One notable recent example is Ariana Reines's *Coeur de Lion* (New York: Fence Books, 2011), an epistolary series of poems written as emails to a lover.

45. See John Mullan, *Anonymity: A Secret History of English Literature* (Princeton, N.J.: Princeton University Press, 2008) and Anne Ferry, "'Anonymity': The Literary History of a Word," *New Literary History* 33, no. 2 (Spring 2002): 193–214.

46. On the social and political logics of strategic anonymity, see Steve Matthews, "Anonymity and the Social Self," *American Philosophical Quarterly* 47, no. 4 (October 2010): 351–63.

47. See Sharon O'Dair, "Laboring in Anonymity," *symplokē* 16, no. 1/2 (2008): 7–19.

48. See Virginia Woolf, "'Anon' and 'The Reader': Virginia Woolf's Last Essays," ed. Brenda R. Silver, *Twentieth Century Literature* 25, no. 3/4 (Autumn–Winter 1979): 356–441.

4. IMPROVISATION

1. On voice in poetry, see David Nowell-Smith, *On Voice in Poetry: The Work of Animation* (New York: Palgrave Macmillan, 2015), and Lesley Wheeler, *Voicing American Poetry: Sound and Performance from the 1920s to the Present* (Ithaca, N.Y.: Cornell University Press, 2008).

2. Wheeler, *Voicing American Poetry*, 2.

3. *Oxford English Dictionary*, "improvisation."

4. Paul F. Berliner, *Thinking in Jazz: The Infinite Art of Improvisation* (Chicago: University of Chicago Press, 1994), 95–105.

5. Karl Coulthard writes, "Jazz, in its earliest incarnations before the music became available on phonograph recordings, generally existed only during the instant of its

performance, a performance characterized by spontaneity and improvisation" ("Looking for the Band: Walter Benjamin and the Mechanical Reproduction of Jazz," *Critical Studies in Improvisation/Études critiques en improvisation* 3, no. 1 [2007]: 1).

6. Amiri Baraka, "Dope," in *The LeRoi Jones/Amiri Baraka Reader*, ed. William J. Harris (New York: Basic Books, 1999), 263.

7. Meta Du Ewa Jones, "Jazz Prosodies: Orality and Textuality," *Callaloo* 25, no. 1 (2002): 67.

8. Before Columbus Foundation, *Poets Read Their Contemporary Poetry*, Smithsonian Folkways Records, catalog no. FL 9702, 1980, mp3 and pdf, folkways.si.edu/poets-read-their-contemporary-poetry-before-columbus-foundation/album/smithsonian.

9. Meta Du Ewa Jones, *The Muse is Music: Jazz Poetry from the Harlem Renaissance to Spoken Word* (Chicago: University of Chicago Press, 2011), 212.

10. Fred Moten, *In the Break: The Aesthetics of the Black Radical Tradition* (Minneapolis: University of Minnesota Press, 2003), 64, 188.

11. On the change in Baraka's performance style, see William J. Harris, "'How You Sound??': Amiri Baraka Writes Free Jazz," in *Uptown Conversations: The New Jazz Studies*, ed. Robert G. O'Meally et al. (New York: Columbia University Press, 2004), 312–25. As Tyler Hoffman notes, Ginsberg recalls beginning to adopt a less academic reading style in the mid-1950s, but his style continued to change in the mid-1960s when he added chant, song, and other more outwardly improvisatory techniques to his repertoire (*American Poetry in Performance: From Walt Whitman to Hip Hop* [Ann Arbor: University of Michigan Press, 2011], 251n25).

12. As Jacques Attali writes of musical improvisation, "the most formal order, the most precise and rigorous directing, are masked behind a system evocative of autonomy and chance" (*Noise: The Political Economy of Music*, trans. Brian Massumi [Minneapolis: University of Minnesota Press, 1985], 114–15).

13. David Sterritt, "Revision, Prevision, and the Aura of Improvisatory Art," *The Journal of Aesthetics and Art Criticism* 58, no. 2 (Spring 2000): 164.

14. Drawing on the political rhetoric of freedom and complicity, Melvin James Backstrom warns that "improvising musicians should not imagine themselves free from the structures, contexts, and histories in which they are implicated" ("The Field of Cultural Production and the Limits of Freedom in Improvisation," *Critical Studies in Improvisation/Études critiques en improvisation* 9, no. 1 [2013]: 5).

15. Daniel Belgrad makes the same point about music: "Improvising jazz solos does not consist mainly in inventing new licks, but in stringing together learned licks and references in new and appropriate combinations—just as new usages and phrasings are a greater part of the poet's work than coining neologisms is" (*The Culture of Spontaneity: Improvisation and the Arts in Postwar America* [Chicago: University of Chicago Press, 1998], 180).

16. By "technologies of inscription," I mean any technology that materially alters a substrate in a way that can be interpreted later: this includes writing and painting but also machine-readable magnetic tapes, hard discs, and flash memory. As Alexander

Weheliye writes, "inscription seems to be at the root of any kind of recording: more than recording itself, it seems that sound necessitates a transposition into writing to even register as technology" (*Phonographies: Grooves in Sonic Afro-Modernity* [Durham, N.C.: Duke University Press, 1999], 25).

17. See Raphael Allison, *Bodies on the Line: Performance and the Sixties Poetry Readings* (Iowa City: University of Iowa Press, 2014), 19, and Hoffman, *American Poetry in Performance,* 44.

18. Coulthard, "Looking for the Band," 4.

19. See Weheliye, *Phonographies,* 27–28.

20. Friedrich A. Kittler, *Gramophone, Film, Typewriter,* trans. Geoffrey Winthrop-Young and Michael Wutz (Stanford, Calif.: Stanford University Press, 1999; originally 1986), 190.

21. As Belgrad puts it in *The Culture of Spontaneity,* "The aesthetic of spontaneity emerged in response to the wartime triumph of corporate liberalism and its techniques of 'information management'" (22).

22. Harris, "'How You Sound??'" 313.

23. He first adopted the name Imamu Amear Baraka, eventually dropping the honorific Imamu and changing the first name to Amiri (Lytle Shaw, *Fieldworks: From Place to Site in Postwar Poetics* [Tuscaloosa: University of Alabama Press, 2013], 107).

24. I rely on Harris for the date and location of this reading and cannot confirm it. Correspondence about the first published recording of Baraka's performance, on the 1980 Smithsonian Folkways record, refers to an event in May 1978, but I cannot find other specifics.

25. Harris, "'How You Sound??'" 314, 317.

26. Jacques Coursil, "Hidden Principles of Improvisation," *Sign Systems Studies* 43, no. 2/3 (2015): 226–27.

27. Nathaniel Mackey, *Paracritical Hinge: Essays, Talks, Notes, Interviews* (Madison: University of Wisconsin Press, 2005), 290.

28. Harris, "'How You Sound??'" 313.

29. Attali, *Noise,* 139.

30. LeRoi Jones (Amiri Baraka), *Blues People: Negro Music in White America* (New York: Harper Perennial, 1999; originally 1963), 225.

31. Walton M. Muyumba, *The Shadow and the Act: Black Intellectual Practice, Jazz Improvisation, and Philosophical Pragmatism* (Chicago: University of Chicago Press, 2009), 17, 42.

32. Amiri Baraka, "How You Sound??" in *The LeRoi Jones/Amiri Baraka Reader,* ed. William J. Harris (New York: Basic Books, 1991), 16.

33. Nathaniel Mackey, *Discrepant Engagement: Dissonance, Cross-Culturality, and Experimental Writing* (New York: Cambridge University Press, 1993), 46.

34. Muyumba, *The Shadow and the Act,* 19.

35. Weheliye, *Phonographies,* 7.

36. Backstrom, "The Field of Cultural Production," 3, 7.

37. LeRoi Jones (Amiri Baraka), "The Myth of a 'Negro Literature,'" in *Home: Social Essays* (New York: Akashic Books, 2009), 130.

38. Mackey, *Discrepant Engagement*, 32, 41, 41.

39. Coulthard, "Looking for the Band," 2.

40. Mackey, *Paracritical Hinge*, 187. Mackey later prefers the structural way around improvisation; he calls it "naïve" to see improvisation as "somehow being immediate and instinctual and instantaneous," for improvising musicians "have recourse to a process of selection and combination that draws on a repertoire" (280). These points would seem to contradict his celebration of fugitivity, but such contradictions lay the groundwork for improvisation's political meanings. On fugitive spirit, see also Mackey, *Discrepant Engagement*, 269.

41. Cornell West, "Black Culture and Postmodernism," in *Remaking History*, ed. Barbara Kruger and Phil Mariani (Seattle, Wash.: Bay Press, 1989), 93, 94.

42. Mackey, *Paracritical Hinge*, 290.

43. For two characteristic defenses, see Belgrad, *The Culture of Spontaneity*, 180, and Charles O. Hartman, *Jazz Text: Voice and Improvisation in Poetry, Jazz, and Song* (Princeton, N.J.: Princeton University Press, 1991), 78.

44. Michael Jarrett, *Drifting on a Read: Jazz as a Model for Writing* (Albany: State University of New York Press, 1999), 74.

45. Belgrad, *The Culture of Spontaneity*, 260.

46. Baraka, "How You Sound??" 16.

47. Gérard Genette, *Paratexts: Thresholds of Interpretation*, trans. Jane E. Lewin (New York: Cambridge University Press, 1997), 1, 12.

48. Moten, *In the Break*, 97.

49. Genette, *Paratexts*, 1.

50. Genette, *Paratexts*, 344–45. Regarding authors' performances of their work, Genette recognizes the need for "a whole separate study, for which, fortunately, I lack the means" (370).

51. Genette, *Paratexts*, 344.

52. Genette, *Paratexts*, 10.

53. Gérard Genette, *Seuils* (Paris: Éditions du Seuil, 1987), 15.

54. Before Columbus Foundation, "Dope: Amiri Baraka," side 2, track 5 (mp3).

55. Meta Du Ewa Jones, "Politics, Process & (Jazz) Performance: Amiri Baraka's 'It's Nation Time,'" *African American Review* 37, no. 2–3 (Summer—Autumn 2003): 249.

56. See Harris, "How You Sound??"

57. Harris cites the version of Baraka's "Dope" performance on *Every Tone a Testimony: An African American Aural History*, Smithsonian Folkways Records, catalog no. SFW CD 47003, 2001, disc 2, track 20, 2001. Another truncated version of this recording appears on the companion audio CD to an anthology Harris coedited, *Call & Response: The Riverside Anthology of the African American Literary Tradition* (New York: Houghton Mifflin, 1998). The plurality of sources for this recording

underscores the ease of losing paratexts within today's infrastructures for preserving poetic audio.

58. Baraka, "Dope," 263.

59. "Amiri Baraka, Diane diPrima and Robert Duncan reading, July 1978," *Naropa Poetics Audio Archives*, Internet Archive, embedded audio, 1:21:40, archive.org/details /Amiri_Baraka__Diane_diPrima_and_Robert_D_78P108.

60. Baraka, "Dope," 263.

61. Baraka, "Dope," 263–64.

62. Baraka, "Dope," 264.

63. Moten, *In the Break*, 63.

64. Amiri Baraka, "Western Front," in Harris, *The LeRoi Jones/Amiri Baraka Reader*, 216.

65. "Amiri Baraka, Diane diPrima and Robert Duncan reading, July 1978."

66. Jones, "Politics, Process & (Jazz) Performance," 251.

67. Genette, *Paratexts*, 1.

68. Amiri Baraka, "Wailers," *Callaloo* 23, special issue [dedicated to Larry Neal] (Winter 1985): 256.

69. See Amiri Baraka, "Amiri Baraka—Wailers," YouTube video, 4:37, youtube .com/watch?v=NoknZIf3HLs.

70. Amiri Baraka, "Amiri Baraka: Poems Recorded at UCLA," May 18, 1983, *Pacifica Radio Archives*, PRA Archive #KZ4239, 21:27, archive.org/details/pra-KZ4239.

71. See "Simon Ortiz and Amiri Baraka reading, July, 1984," July 28, 1984, *Naropa Poetics Audio Archives*, embedded audio, 21:43, archive.org/details/Simon_Ortiz_ Amiri_Baraka_reading_July_1984_84P060.

72. See LeRoi Jones (Amiri Baraka), "The Changing Same (R&B and New Black Music)," in *Black Music* (New York: Akashic Books, 2010; originally 1968), 175–205.

73. "Allen Ginsberg, Anne Waldman, and Amiri Baraka reading with Steven Taylor song performance, July, 1992," Jack Kerouac School for Disembodied Poetics, Naropa University, Boulder, Colo., July 25, 1992, embedded audio, 1:30:58, archive.org/ details/Allen_Ginsberg__Anne_Waldman__and_Amiri__92P031.

74. Amiri Baraka, "Masked Angel Costume," in *SOS: Poems 1961–2013*, ed. Paul Vangelisti (New York: Grove Press, 2014), 286.

75. Baraka, "Masked Angel Costume," 283.

76. Amiri Baraka, "Why It's Quiet in Some Churches," in Vangelisti, *SOS*, 297.

77. Amiri Baraka, "Funk Lore" and "I Am," in Vangelisti, *SOS*, 370, 312.

78. Baraka comments powerfully on this strategy of historical erasure: "A 'cultureless' people is a people without a memory. No history. This is the best state for slaves; to be objects, just like the rest of massa's possessions" ("The Changing Same," 207).

79. "Allen Ginsberg on William F. Buckley's Firing Line—The Avant Garde," May 7, 1968, YouTube video, 50:39, youtube.com/watch?v=MkoLOj6LwOs. A corrected transcript of this interview is available in the Raymond Danowski Poetry Library at Emory University, and the transcript minus Buckley's introduction is reprinted in

Allen Ginsberg, *Spontaneous Mind: Selected Interviews, 1958–1996*, ed. David Carter (New York: HarperCollins, 2001), 76–102. The video reveals minor inaccuracies in both transcripts, and my quotations refer to the video. *Spontaneous Mind* will hereafter be cited as *SM*.

80. "Excerpts from the Boston Trial of *Naked Lunch*," in William S. Burroughs, *Naked Lunch* (New York: Grove/Atlantic, 1990; originally 1959), xix. On Ginsberg's appearance at the trial, see Michael Schumacher, *Dharma Lion: A Biography of Allen Ginsberg*, expanded ed. (Minneapolis: University of Minnesota Press, 2016), 416–17.

81. The government's authority to sanction profanity on television stems from U.S. Code Title 18, §1464, enacted in 1948, which forbids broadcasting "any obscene, indecent, or profane language by means of radio communication." In the 1960s, the Federal Communications Commission (FCC) occasionally sent warning letters, but broadcasters self-policed. The FCC issued its first indecency fine in 1970 when a Philadelphia radio station aired an expletive-laden interview with Jerry Garcia (Lili Levi, "The FCC's Regulation of Indecency," *First Reports* 7, no. 1 [April 2008]: 10–11).

82. Wheeler, *Voicing American Poetry*, 12.

83. Allen Ginsberg, *Planet News 1961–1967* (San Francisco: City Lights, 1968), 12.

84. Ginsberg, *Planet News*, 108, 27.

85. Allen Ginsberg, *The Fall of America: Poems of These States 1965–1971* (San Francisco: City Lights, 1972), 30, 29, 1.

86. Ginsberg, *The Fall of America*, 1.

87. Quoted in Alice Echols, *Shaky Ground: The Sixties and Its Aftershocks* (New York: Columbia University Press, 2002), 26.

88. Jacques Derrida, *Margins of Philosophy*, trans. Alan Bass (Chicago: University of Chicago Press, 1982; originally 1972), 8.

89. N. Katherine Hayles, "Voices Out of Bodies, Bodies Out of Voices: Audiotape and the Production of Subjectivity," in *Sound States: Innovative Poetics and Acoustical Technologies*, ed. Adalaide Kirby Morris (Chapel Hill: University of North Carolina Press, 1997), 77.

90. Michael Aldrich, Edward Kissam, and Nancy Blecker conducted the interview on November 26, 1968, in Cherry Valley, N.Y. Mark Robison wrote an introduction to the interview and published a limited edition of 2,000 copies in 1971. The interview was reprinted first in *Composed on the Tongue: Literary Conversations, 1967–1977*, ed. Donald Allen (San Francisco: Grey Fox, 1980), and again in *SM*. See also Allen Ginsberg, *Improvised Poetics*, ed. Mark Robison (Buffalo, N.Y.: Anonym Press, 1971).

91. Ginsberg, *Composed on the Tongue*, 38.

92. The reference is to William Blake, *Songs of Innocence and Experience*, tuned by Allen Ginsberg, New York, December 15, 1969: (New York: MGM Records, 1970; MGM-Verve FTS-3083), side 1, band 4. The full record is available on *PennSound*: writing.upenn.edu/pennsound/x/Ginsberg-Blake.php.

93. Jack Kerouac, "The Essentials of Spontaneous Prose," in *The Portable Beat Reader*, ed. Ann Charters (New York: Penguin, 1992), 57. An earlier version of the essay first appeared in *Black Mountain Review* 7 (Autumn 1957): 226–28.

94. William Wordsworth, "Preface to *Lyrical Ballads*," in *William Wordsworth*, ed. Stephen Gill, 21st-Century Oxford Authors (New York: Oxford University Press, 2010), 65.

95. William S. Burroughs, "Origin and Theory of the Tape Cut-Ups," Jack Kerouac School of Disembodied Poetics, Naropa Institute, Boulder, Colorado, April 20, 1976, YouTube video, 3:29, youtube.com/watch?v=gKvL-V8Fu_U. In this lecture Burroughs suggests "that perhaps when you cut into the present [with the cut-up method] the future leaks out."

96. Belgrad, *The Culture of Spontaneity*, 197.

97. Jarrett, *Drifting on a Read*, 78.

98. The slippage between "where" and "when" here again dramatizes the becoming-space of time central to the inscriptive process that renders a "when" of enunciation as a "where" on a substrate of inscription (Ginsberg, *SM*, 125).

99. By the 1870s, researchers had used photoengraving to transfer phonautographs onto surfaces from which they could be replayed, but Scott's earliest recordings remained unheard until 2008, when scientists at the Lawrence Berkeley National Laboratory took digital photographs of a phonautograph drum and used a computer to convert these images to audible waveforms (Jody Rosen, "Researchers Play Tune Recorded before Edison," *New York Times*, March 27, 2008, nytimes.com/2008/03/27/arts/27soun.html).

100. Édouard-Léon Scott de Martinville, "Principles of Phonautography," in *The Phonautographic Manuscripts of Édouard-Léon Scott de Martinville*, ed. and trans. Patrick Feaster, FirstSounds, March 2010, firstsounds.org/publications/articles/Phonautographic-Manuscripts.pdf.

101. Jonathan Sterne, *The Audible Past: Cultural Origins of Sound Reproduction* (Durham, N.C.: Duke University Press, 2003), 46.

102. Ginsberg's interest in mantra chanting began during his 1962–63 trip to India. He reports beginning to chant during readings in 1964 and intensifying his practice in 1968 ("The Vomit of a Mad Tyger," *Lion's Roar: Buddhist Wisdom for Our Time*, April 2, 2015, lionsroar.com/the-vomit-of-a-mad-tyger/). See also Bill Morgan, *I Celebrate Myself: The Somewhat Private Life of Allen Ginsberg* (New York: Viking, 2006), 392.

103. This movement away from the verbal alters the voice's coordination of alphabetic and acoustic inscription, interrupting a mise-en-abyme that works between these two technologies. The inscriptive account of improvisation emerges from this interplay between paper-based scripting (from which the performer must depart in order to improvise) and more indexical modes of inscription (i.e., audio or video recording) that capture the performance.

104. "Allen Ginsberg on William F. Buckley's Firing Line."

105. Allison, *Bodies on the Line,* 4–5.

106. "Writing and Poetics Performance with Allen Ginsberg, Anne Waldman, Steven Taylor, and Amiri Baraka," tape 1, Jack Kerouac School of Disembodied Poetics, Naropa Institute, Boulder, Colorado, July 25, 1992, embedded audio, 1:30:58, archive.org/details/Allen_Ginsberg__Anne_Waldman__and_Amiri__92P031.

107. Coursil, "Hidden Principles of Improvisation," 232.

108. *Naropa Poetics Audio Archives,* archive.org/details/naropa.

109. "Allen Ginsberg and Philip Whalen reading, June, 1975," Jack Kerouac School of Disembodied Poetics, Naropa Institute, Boulder, Colorado, 1975, embedded audio, 3:30, archive.org/details/Allen_Ginsberg_and_Philip_Whalen_reading_June_1975_75P108.

110. The Naropa University Archives also specifies the date of the reading as June 18, 1975, cdm16621.contentdm.oclc.org/cdm/singleitem/collection/p16621coll1/id/194/rec/8.

111. Emily Morris, Technical Services Librarian, Allen Ginsberg Library, Naropa University, personal email to the author, March 3, 2017.

112. "Allen Ginsberg and Philip Whalen reading, June, 1975."

113. Charles Bernstein, "PennSound Manifesto," *PennSound,* University of Pennsylvania, writing.upenn.edu/pennsound/manifesto.php.

114. Bernstein, "PennSound Manifesto." The same six points are elaborated in Charles Bernstein, "Making Audio Visible: The Lessons of Visual Language of the Textualization of Sound," *Text* 16 (2006): 277–89.

115. Bernstein, "PennSound Manifesto."

116. As Jonathan Sterne writes, the MP3 is not open source but "a proprietary standard that brings in hundreds of millions of euros each year for the companies that hold the rights to it," even if it "feels like an open standard to end-users" (*MP3: The Meaning of a Format* [Durham, N.C.: Duke University Press, 2012], 26).

117. "Robert Frost," ed. Chris Mustazza, *PennSound,* writing.upenn.edu/pennsound/x/Frost.php.

118. Stranger still, the site dates these Frost recordings to "May 5, 1933 and October 24, 1934," but commentary on the same page claims the recordings "were made . . . on January 16, 1934 and March 10, 1934."

119. Bernstein, "PennSound Manifesto."

120. Bernstein, "Making Audio Visible," 283, 284.

121. Allison, *Bodies on the Line,* 37.

122. See Charles Bernstein, *Content's Dream: Essays 1975–1984,* Avant-Garde and Modernism Studies (Evanston, Ill.: Northwestern University Press, 2001; originally 1986), 246–48, and Bernstein, "Seminary Co-op Books (*Front Table*) Interview," Seminary Co-op Bookstore, Chicago, February, 1999, *Electronic Poetry Center,* writing.upenn.edu/epc/authors/bernstein/interviews/SemCoop.html.

CONCLUSION

1. Most notably, Jessica Pressman reads both the Poundstone and the YHCHI pieces in *Digital Modernism: Making It New in New Media* (New York: Oxford University Press, 2014). On the Poundstone piece, see also Jessica Pressman et al., *Reading Project: A Collaborative Analysis of William Poundstone's Project for Tachistoscope {Bottomless Pit}* (Iowa City: University of Iowa Press, 2015).

2. All three poems can be viewed on the web. *Dakota* and other works by YHCHI are available at yhchang.com. *Project for Tachistoscope* appears in volume 1 of the *Electronic Literature Collection,* collection.eliterature.org. Tan Lin's piece is at www.supercentral.org/cookingsystem.

3. For an account of the aesthetic and political stakes of poetic objecthood in modernist writing and after, see Lisa Siraganian, *Modernism's Other Work: The Art Object's Political Life* (New York: Oxford University Press, 2012).

4. Michael Fried, "Art and Objecthood" (1967), in *Art and Objecthood: Essays and Reviews* (Chicago: University of Chicago Press, 1998), 150. Hereafter, this essay will be cited parenthetically in the text.

5. See Lisa Gitelman, *Paper Knowledge: Toward a Media History of Documents* (Durham, N.C.: Duke University Press, 2014), and Jonathan Senchyne, *Intimate Paper and Materiality of Early American Literature,* forthcoming from the University of Massachusetts Press.

6. Among other critics, Pressman has made this connection. She notes that "the translation of computer code into human language produces electronic literature." ("Reading the Code Between the Words: The Role of Translation in Young-hae Chang Heavy Industries's *Nippon*," *Dichtung-Digital* 37 [2007], dichtung-digital.de/2007/pressman.htm).

7. Jonathan Culler, *Theory of the Lyric* (Cambridge, Mass.: Harvard University Press, 2015), 226. In a narrower but earlier version of this point, Sharon Cameron identifies in Emily Dickinson's poetry a "resolute departure from temporal order" (*Lyric Time: Dickinson and the Limits of Genre* [Baltimore, Md.: Johns Hopkins University Press, 1979], 1).

8. Pressman, *Digital Modernism,* 56.

9. This speed is one important way that the presentness of these poems differs from that of minimalist sculpture. The *now* that such sculptures invoke is expansive, geological in scale, a "long now," so to speak, but the computer poet's sense of temporal presentness seems faster, more instantaneous, unfolding at the speed of light rather than stone. Blink and you miss it.

10. Pressman, *Digital Modernism,* 63.

Index

SETH PERLOW is assistant teaching professor of English at Georgetown University. He edited Gertrude Stein's *Tender Buttons: The Corrected Centennial Edition.*